# Walking to the Sun

Tom
Haines

# Walking to the Sun

A Journey
through America's
Energy Landscapes

❋

ForeEdge

ForeEdge

An imprint of University Press of New England

www.upne.com

© 2018 Tom Haines

All rights reserved

Manufactured in the United States of America

Designed by Eric M. Brooks

Typeset in Whitman by Passumpsic Publishing

Maps by Matthew James Wimett

For permission to reproduce any of the material
in this book, contact Permissions, University Press of
New England, One Court Street, Suite 250, Lebanon
NH 03766; or visit www.upne.com

Library of Congress Cataloging-in-Publication Data
available upon request

Paperback ISBN: 978-1-5126-0095-7

Ebook ISBN: 978-1-5126-0316-3

5  4  3  2  1

*For*

*Shilla, Abdo, & Bontu,*

*Luca & Colette*

❊

*Not till we are lost, in other*
*words not till we have lost the world,*
*do we begin to find ourselves,*
*and realize where we are and the*
*infinite extent of our relations.*

❋

HENRY DAVID
THOREAU

# Contents

Preface
xi

## Part 1 **Into Today**

CHAPTER 1 Boom Time Tornado
3

CHAPTER 2 States of Combustion
55

CHAPTER 3 Black Thunder Bust
91

## Part 2 **On to Tomorrow**

CHAPTER 4 Idle Currents
127

CHAPTER 5 Turning Time
151

CHAPTER 6 Whether to Burn
185

Acknowledgments
207

# Preface

I grew up in the 1970s in a small town on the Ohio River, seven miles downstream from the city of Pittsburgh. Our house sat at the end of a dead-end street. The backyard had towering oak trees, and, beyond, there was a steep ravine. Some of my earliest and strongest memories of leaving the house involve cutting across the lawn and heading down into the woods, deep, all the way to a stream. It was only a few feet across. My friends and I spent hours there, stacking little rocks to make pools and waterfalls.

Less than a mile in the other direction was Neville Island. During my childhood, barges churning up the Ohio River from the Mississippi unloaded coal on Neville Island for conversion to coke, a fuel that fires furnaces for manufacturing steel. On summer nights, when I lay in my bedroom with the window open, I could hear the grinds and whines of the coke plant, one small piece of the vast system that has insulated humans from nature.

Decades later, as a journalist reporting in dozens of countries, I was drawn to parts of the world where industrialization had not yet arrived. There families gather around open fires and sleep in huts. I think of villagers I met in the Great Rift Valley of Ethiopia: Gebi and Halima and their children, Shilla, Bontu, and Abdo. Many of the six billion people in developing parts of the planet ache for the comfort and convenience of the industrial world. Yet during the decades of my life, the tension between nature and industry has become ever more dire, as the burning of fossil fuels to keep everything running warms the planet and accelerates extreme climate change. Rising temperatures are melting polar ice, lifting seas, strengthening storms.

On a December day in 2013, I watched a man bolt a new natural gas furnace to the basement wall of a house in New Hampshire, where I had settled with my wife and two young children. As the machine kicked on, burning natural gas to heat water to warm our home, I could not fathom the scale of my own consumption. I knew that such systems must shift from fossil fuels to sources that are renewable and carbon-free. But staring at the narrow pipe that ran from the furnace through the basement wall and into the world beyond, I felt a deep dislocation from any larger understanding: can we get from today to tomorrow?

I decided then to go to the source. I plotted a loose route that would form a reckoning and an exploration: three journeys across landscapes of fossil fuel and three across landscapes of things renewable and carbon-free. Oil, gas, and coal. Water, wind, and sun. In each place, I would move on foot. An animal, vulnerable and alert.

*Walking to the Sun* tells the story of this journey that we all are on.

# Part 1

# Into Today

# Chapter 1
# Boom Time Tornado

Sand Creek Road, as McKenzie County Road 2 is also known, turns from asphalt to hard-packed dirt just past the small bay at Tobacco Garden and climbs steeply to a ridge before tracing the rugged terrain south of Lake Sakakawea. The lake is really a reservoir, and it was formed in the early 1950s, after the construction of the Garrison Dam blocked the flow of the Missouri River to control seasonal flooding and generate hydroelectric energy for farming communities planted across the North Dakota prairie. Prior to the construction of the dam, more than a thousand people living alongside the river on land of the Mandan, Hidatsa, and Arikara Nation were forced to move to a settlement called New Town. The land forming the reservoir's new shore—previously more of a midpoint in what had been higher hills—had been carved during thousands of years by creeks and streams draining into the river, and today trees are few and stiff grass roots in dry earth. Two miles east of Tobacco Garden, Sand Creek Road descends into a shallow canyon, and the iron bars of a cattle guard mark the point at which it enters a fenced pasture.

On the spring day I was walking along the road, there stood, twenty feet on the other side of the iron bars, a black bull. He weighed a thousand pounds and more. The bull's head bowed against the earth, and his body swayed to the cadence of his munching mouth. It was nearly noon, and a weathered piece of wood holding up a nearby section of barbed-wire fence rocked against steady gusts that gave the valley a feel of constant action, though no one else was there. I shrugged off my backpack and sat on the ground to drink from my water bottle. The bull began moving toward me. His inhalations and exhalations had the force and volume of a well-powered vacuum cleaner. He stopped just short of the cattle guard and snorted and bellowed, agitated by my arrival.

It was May 2014, and thousands of square miles of western North Dakota were overrun in an unprecedented oil boom. News reports told stories of overnight millionaires and broken dreams, of so much oil flowing out of the prairie that it was changing international politics, and of gas flares burning bright enough to be seen from space. The Bakken oil field, as it had been named, was the latest place to be transformed during the century that crude oil had become a favorite for making fuel for everything from automobiles to airplanes.

I had only ever encountered that fuel at the end of a nozzle, as I filled my tank at gas stations while running errands, driving to work, or off on a family vacation. What could the controlled chaos of the harvest be like up close, with so many people scrambling to claim more energy? The North Dakota boom was as good a place as any to begin to encounter at ground level the source of my consumption.

The center of the drilling action that summer would be in the eastern half of McKenzie County, where dozens of rigs rose into the sky, and that was my destination by day's end. But I wanted to arrive slowly, to feel the North Dakota earth as it had been before the boom. So, when I had plotted my walk that spring back home in New Hampshire, zooming in and out on a satellite view of McKenzie County on Google Earth, I'd been drawn to the line that marked County Road 2 and the rugged terrain through which it wove south of Lake Sakakawea. The ten-mile route cut through land that had not yet been heavily drilled and then crested onto open prairie and the thick of the boom near the town of Charlson. Sand Creek Road, in other words, offered a chance for me to clear my mind and find my footing.

Though I was hungry, I wanted to eat enough only to keep going. As I sat across from the bull, I dug into a pouch for a granola bar. I thought I'd be able to wait the bull out. Perhaps he would wander off toward the far end of the pasture. Then I heard, coming from behind me, more bellows. I turned to see a second bull, also a thousand pounds and more, lumbering toward me. No cattle guard or fence separated us. I must have crossed another cattle guard while walking earlier, but I had not been paying proper attention, as it was only my first day.

I scrambled across the road, dropped to my belly, and soldier-crawled beneath the barbed-wire fence into a third pasture, thick with dried cow patties. I rose, dusted off, and began to walk toward a ranch house set back in a draw. When I got about fifty feet away, a woman came outside. She had cropped hair and wore a billowy T-shirt, jeans, and rubber mud boots, which she planted firmly on the ground. She did not return my approaching wave, and I'm pretty sure she did not respond when I called "hello." There was not another home for miles around. I apologized for walking up unannounced and explained my predicament with the road-blocking bulls. The woman stared into the half distance and said that the bulls were hers. She said she wouldn't walk past them if she were me. "You never know what bulls will do," she said.

I asked if she could give me a ride in her pickup truck to pass the bulls. She shook her head and nodded toward the barn next to the house. "I have to check on cows." She said that if I was determined to keep walking up the road I better stick to the fence line and move slowly. "Bulls are strong and unpredictable," she said. Then, in what seemed her first kindness, she added, "Now, I'm not trying to scare you."

But she wasn't talking about bulls anymore.

"There are mountain lions out here," she said, "and coyotes. There was just a mountain lion shot near here last week. Are you going to camp out here tonight?"

"I'm walking six or seven more miles this afternoon," I said. "I was hoping to pitch my tent alongside the road."

She stared past me. "You're on your own," she said.

Unsure, and alone again, I walked back across the pasture to the road. The first bull was still standing near the cattle guard. But the second bull had cleared out of the area where I'd left my backpack. I crawled back under the barbed-wire fence. Only three vehicles had passed as I was walking that morning, and I was not hopeful another would come along soon. I idled in indecision, and then I saw a cloud of dust in the direction I planned to head and, at its center, a white pickup truck motoring toward me. The big king cab pulled to a stop, and the passenger rolled down his window. A tattooed arm emerged from the air-conditioning. The driver looked across the front seat and laughed.

"What, you don't want to carry your big red backpack past those bulls?" There must have been more bulls farther into the pasture. I tried to look like I knew what I was doing. He spun the truck around and then lowered his window. "Hop on the back."

I threw my pack into the truck bed and climbed on top of a shiny steel toolbox. As we crossed the cattle guard and began to accelerate, the first bull lowered his head, bowing before the big machine. After a quarter-mile the truck tires rumbled across the iron bars of another cattle guard. The driver shouted out an offer to keep going.

"There's a big hill up there," he said.

"I'll just walk," I called back.

He stopped the truck, and I jumped to the ground.

Just three hundred years ago, daily survival still meant using individual effort to gather fuel to make it through the day. A colonial family, with several kids and a pair of oxen, could harness the energy equivalent to three horsepower. That amount of energy, which came mostly from muscle, had to chop wood and plant crops and prepare food, and there was a known intimacy in the transaction between a person and the earth.

Back home in New Hampshire, I lived deep inside the industrial world: my home heated through winter, my refrigerator chilling food from another hemisphere, two cars in the garage waiting to carry me wherever I needed to go with the simple push of a pedal. My family, as is typical for our time, consumed more than a hundred times the energy of the colonial family, yet my relationship to the source was distant. I worked in an office and sent electronic payments for power I seldom saw, as it arrived in kilowatts through the grid or poured unseen from pump to fuel tank.

My walk into the prairie oil field was a first attempt to overcome that separation. I had chosen to walk because I wanted to measure the exhaustion of my effort and encounter the oil harvest not as a product of the system it supported but as an animal exposed, part of nature again. I wanted to break through the over-insulation of my life back home, to see and smell and touch the source of my survival, and to begin to know the scale at which I live in the early edge of the twenty-first century.

I followed Sand Creek Road another mile, and the shallow valley narrowed. A steep rock wall framed its northern edge. I approached a dense stand of trees, and I looked at the ridge and wondered: mountain lion? My pulse quickened, but only briefly. After a few hundred more yards, the road turned a tight corner and entered open country, climbing the steep hill the driver of the pickup truck had mentioned. In a ditch just off the road to the right, iron bars had been fashioned into two crosses, and each was capped with a white hard hat. Black letters on one hard hat spelled "Porter" and "Dad." At the base of the cross, a steel-toed work boot and an empty bottle of Bud Light had been left in memoriam. An arrow had also been stuck in the ground nearby. Black letters on the second hard hat spelled "Seymour" and "Dad." Another beer bottle, a cigarette, and a withered orange lay at the feet of this cross. Brittle stalks of grass bent from their perches in the dirt.

Thirty minutes of hard walking brought me to the top of the hill. There I saw no one, but the solitude was gone in an instant, as I'd entered a land of moving machines. A mile or more ahead, pickup trucks shuttled between oil wells and tanker trailers heaved beneath liquid loads. To my left the single horsehead pump of a well rose and fell at a steady rhythm. Its iron legs creaked as unseen oil was piped to nearby tanks. The well pad was fenced, and signs noted that the well was named the Pittsburgh 1-3H, operated by Newfield Production Company. Another sign warned,

1. HARD HATS REQUIRED!
2. SAFETY GLASSES REQUIRED!
3. STEEL TOE BOOTS REQUIRED!
4. NOISE PROTECTION REQUIRED!
5. NO SMOKING OR OPEN FLAMES!
6. AUTHORIZED PERSONNEL ONLY!
7. NO FIREARMS!
8. NO ALCOHOL!

I stumbled on, tired from the climb but also bewildered by the uncertain order of the scene all around, and soon I came to another well pad, which had three pumps, side by side. They stood thirty feet tall or more, and from my angle, on foot, each pump did look like a horse rearing onto its hind legs and dropping down again. These wells were the Thompson 21-11SEH, 21-11SH, and 21-11SWH, owned and operated by XTO, a subsidiary of Exxon.

A double tanker trailer rumbled up a side road, turned on to County Road 2 heading in my direction, toward Charlson, and passed within a few feet of me. I stood on the edge, bathed in dust. Another white pickup truck skidded to a stop and another passenger window descended. Two big men in the air-conditioning looked out at me in wonder. They were pavers from Oklahoma, hustling around the farms-turned-oil-field to drum up extra business between well-pad work. We chatted briefly, but they seemed little concerned with what I was doing there. Before the window went back up, the passenger paused. "Take care of yourself," he told me. "Drink plenty of water."

I had one bottle left. The pavers didn't have any extra in their king cab, so I dropped my pack and walked toward a backhoe that was leveling earth. A young man monitoring the work said he didn't have any extra water either. He said he was born and raised in Johnson's Corner, about twenty miles south. I asked him about all the action. "I don't like it," he said. "Brings too many people in."

I walked on, unmoored from the purpose of the oil patch. After six hours and ten miles of walking in eighty-degree heat, I was invigorated in a way I was not used to at home. I stalled above an empty scrape of earth that had been cleared as a turnaround for trucks, then sat down and stared past my boots at the ground. A twig leaned against a pebble. There was no life. The patch of land was neither of the prairie nor anywhere else.

It was already evening, and I headed into the pasture alongside the road to camp.

The land sloped downward before rising again, and in the swale

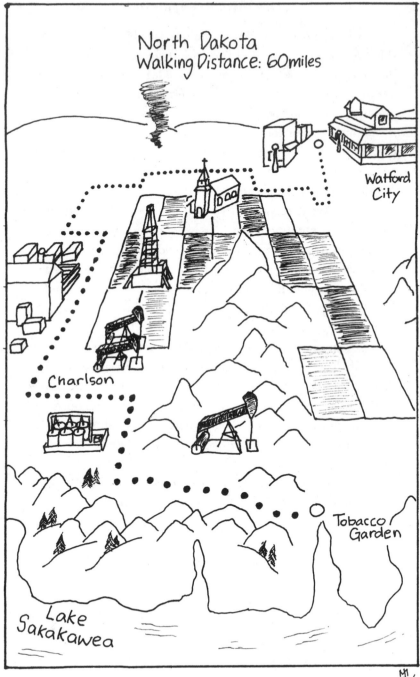

I hustled to pitch my tent. Afterward, happy to be free of forty-five pounds of gear, I decided to stroll around. I followed the slope back toward the road, and under some power lines I noticed two yellow poles. A sign posted on the top of one pole read, "Danger: High pressure gas pipeline." Then, one on another pole nearby read, "Warning. Poison Gas/$H_2$S May Be Present." I knew that hydrogen sulfide can be both impossible to smell and deadly, so I carried my tent farther into the prairie, up the next rise, then down the other side. The pasture was knee-high grass for miles to the south, and in the distance I could see seeded fields and a few ranch and farm buildings near Charlson. I cooked dinner, then climbed into my tent an hour before sundown.

There was no more traffic along Sand Creek Road, as trucks had finished their rounds to the wells, and I was the only human for miles around. I arranged my water bottle, a flashlight, and notebooks at one end of my tent. I scrunched clothes into a sack for a pillow. Perched on the prairie at the end of the day, I had a sense of arrival, but I knew that was only a beginning. The next morning I would walk into Charlson and the action all around. As a journalist, I was confident I would see much of how the oil field worked. As a lone wanderer, I hoped to find understanding about my involvement in the system.

I crawled out of my sleeping bag at dusk to go to the bathroom before sleeping. Quieting gusts shook the tethered tent. Songbirds darted low. The sweep of cobalt blue overhead shaded toward black, as everything close drew inward. But on the horizon, toward Charlson, shocks of bright light stretched the line of sight. The fields had disappeared in darkness, and three miles to the south, flares burned natural gas that was rising quickly with oil from deep-drilled wells. The wells had been indiscernible from this distance in the daylight. Yet at night they defined the terrain. I could not hear the audible roar of each flare as fuel met flame, and in that moment the brilliant light formed a benevolent border between buried rock and shifting sky. There came a feeling of floating. I stared at the horizon and counted: 1 flare, 2 flares, 3 flares . . . 4, 5, 6 . . . 7, 8, 9 . . . 10, 11, 12 . . . 13, 14, 15 . . . 16, 17, 18. . . . Orange and yellow flames leaped above unseen pipes, and the constant wildness of the fire seemed an organic part of the place, as if meant by nature to exist.

*

By midmorning two days later, in a pasture ten miles south, dust blew up from brown earth, and two high school boys, sturdy kids named Shane and Justin, rested on the tailgate of a pickup truck parked on the prairie. The boys had just finished the forceful separation of more than a hundred calves from their mothers, with the cows moved out of a makeshift pen to a larger pasture. The penned calves, born on the range weeks earlier, would spend the summer and early fall grazing grasslands before being shipped to commercial lots and fattened up to feed the appetites of people far away. On this unexpected morning of shifting sky and sudden gusts, the commoditization began.

Doug Olson, who owned the calves, set four long irons with his family's brand in the flame of a propane torch. Most of the cows had wandered over a hill to wait, but one cow maneuvered behind the crowd of people—several generations of several local families who had gathered for the branding—and she wailed at the calves, who stumbled around the pen as though they no longer understood the earth beneath their hooves. When the iron brands pulsed orange, the high school kids jumped down and lined up outside the corral. Marco Pelton, whose father had taught agriculture tech at the high school in Watford City, the seat of McKenzie County, sauntered up to me and asked if I was going to get to work too. It was a fair question. I had shown up unannounced that morning with Leif Jellesed, a rancher and farmer I'd met the day before when I walked into Charlson after waking at my campsite at the edge, and he'd invited me to join him at the branding. I was curious to know what the oil boom meant for those who called the prairie home. But as Shane, Justin, and the others got to work, my role seemed uncertain: observer or participant?

Two older men on horseback, one stout and somber beneath his gray beard, the other lean like his rope, rode into the corral. The lean cowboy roped a calf, and his horse dragged it toward the teenagers. One of them grabbed the taut rope as it slid by and jerked it high. The calf spun in the air and hung there, briefly, before landing on its back. Shouts and whistles and bellows coursed all around—so many people doing so many jobs—but the young man stayed focused. He sat and held the calf's hind legs as another teenager pinned the calf's front legs and head. If the boys strained, their muscles did not show it, as they were in their animal prime.

Ray Gilstad, who fifty years before had won the state's bareback riding championship, strode up with a glowing iron, planted one boot on the calf's thick side for control, and pressed the brand into the flesh. The calf braced as smoke singed the air. Beau Wisness, a bull of a man, bent low and stuck a long needle in the calf's side, then squeezed a syringe full of vaccine that prevents a disease called black leg. The calf kicked. Leif, who runs a herd of his own cattle several miles north and east, bent in close with a quick tug on the calf's scrotum and flick of a razor blade to remove the testicles. The calf huffed against the dirt. The boys climbed off the calf's legs, and the animal stumbled upright, its wet eyes darting without direction. It staggered for a few steps, then bolted toward open pasture.

I took a spot in line behind the teenagers. When it came my turn to flip a passing calf, I grabbed the rope and yanked, but not hard enough. The calf rose into the air but made only a quarter-turn and landed on its side, legs still scrambling for traction. I leaped to catch the calf's hind left leg, my forty-six-year-old frame arching against the effort. I dug my boot into its other hind leg and pulled hard, trying to immobilize the animal. A blond boy with the sleeves cut from his T-shirt moved in quickly and sat on the calf's head, holding the front legs with one hand. Marco leaned down and suggested that I cover the calf's rear end, which was pointed right at me, with my other boot. Branding can be so upsetting to a calf, Marco said, that it can lose bowel control, and my closeness meant exposure. For more than an hour I wrestled one calf after another, and as I tried to pin a particularly strong calf to the ground, Doug strolled over. Mounds of muscle in the calf's leg twitched and tightened beneath my grip. Course bristles of the hide scraped against my palms. The calf jerked and braced as the men leaned in to burn, and inject, and slice. I didn't understand that holding with easy pressure would calm the calf, so I struggled too hard to keep it pinned. Doug looked toward me and said, with a laugh, "You're not going to be able to walk so far tomorrow."

When I'd arrived on foot at Leif's ranch the day before, it was clear that I was moving at a speed long since abandoned on the prairie. The route from my campsite overlook toward Charlson had passed an oil well–drilling rig, a city unto itself that loomed twenty stories tall, and the pulse of traffic—tankers and tractor trailers, septic trucks servicing worker camps, and one-ton pickups carrying people to and fro—became more constant on that last mile of Sand Creek Road. The road had angled south, away from Lake Sakakawea, and the prairie opened toward the horizon, corduroy rows of freshly tilled fields as far as I could see. I had spotted Leif's new metal equipment shed, large enough to house a small airplane, from a mile or so away, and when I turned down the long driveway and looked in the open bay doors, I found Leif inside.

Leif was wearing a baseball cap, T-shirt, and jeans, perfect attire for an afternoon of tinkering and tending to the two-thousand-acre farm on which he plants spring wheat, flax, canola, and more.

Two hired hands, farmers from South Africa come to make more money than they could at home on farms of their own, were out in the fields, sowing even rows with seeds. But their day, which began with the punch of a clock that morning at 6:51 a.m. and ended near midnight, would be spent seated in air-conditioned tractors. The springtime ritual of branding calves by hand was an exception in the world of work. Most planting of crops and herding of cattle is done by a few people driving big machines. "There is no labor in farming anymore," Leif told me.

In 1915, just down the road from Leif's new shed, his grandfather had built a red wooden barn. He'd hauled in wood from Tioga, across the Missouri River, and family and neighbors, others newly arrived in the prairie, had helped hoist the frame. As the years passed and the farming community took root, Leif's father added a concrete foundation beneath the barn, and the town became home to hundreds of residents. But in recent decades, as industrialization of U.S. agriculture accelerated, there was less work to be done, and many moved away. By the time I arrived, dozens of smaller farms had consolidated into the hands of a few, and Charlson was home to just four families, including Leif's. The red barn still stood, but its second floor, once a granary to keep food for livestock through the long winter, was mostly empty. Old saddles hung on a wall and pigeon poop lay thick on the floor. Along what once had been Charlson's main street, broken glass hung in dance-hall windows, allowing entry for swallows, who nested above a piano, which stood silent.

Leif and I sat for more than an hour in the shade of his shed, and he told me that since the oil boom hit, beginning in earnest in 2009, eleven wells had been drilled on his land. He owned the mineral rights for some and received payments for surface access on others. The money helped cover the high cost of industrial agriculture. "I couldn't farm the way I'm farming now, without the oil," Leif said. As we talked, tractor trailers hauled rock from a pit northeast of Leif's property. Leif described how the companies install the oil-storage tanks before the wells are even dug. "They know what they're going to get," Leif said. "Not completely, but they know they're going to get oil." My mind raced at the speed and certainty of the enterprise: land that had been prairie grassland for millennia and cultivated for crops for a century was being redefined by the oil harvest in just a few fast years.

Late in the afternoon, Leif's wife, Nancy, rode up on an ATV, and the three of us stood in a small garden near the shed where Nancy had planted rows of asparagus. It was a rare crop, meant for the Jellesed's table. We strolled between the rows, snapping stalks from the soil. Some stalks were a foot long and thin around as a pencil. Others were short and stout. Nancy suggested I eat the asparagus raw. I stood with my boots in the chunky soil, feasting on the taste of sweet earth.

Leif offered me a room in the bunkhouse for the evening. Before he and Nancy headed to their house for dinner, Leif told me he would join Doug Olson the next morning, a Saturday, to brand calves, and that's when he invited me along. The community gathering was an authentic rite of spring. But it was also a cultural relic, a keeping of time in the face of change.

By noon the next day at the Olson Ranch, the first batch of calves had been branded, and Shane and Justin and other teenagers stacked panels from the temporary corral onto a trailer. Cowboys and kids, neighbors and grandparents, piled into pickup trucks and motored along a dirt road past Doug's house and followed a two-lane track to the top of another pasture. Doug's wife, Annette, put pots of sloppy joes and baked beans on a folding table. The cowboys parked the horse trailers in an imperfect circle, and a few dozen people idled about, never too far from the growing buffet. As Doug walked toward a fence, he spotted his cattle dog, which missed the earlier action. "Did we forget you this morning, Duke? Did we?" Doug said, as he bent to rub Duke behind the ears. Duke leaned in. "You're the best dog I ever had," Doug told him. Doug stood and turned to Leif. There were a hundred more calves to brand, and the two men began strategizing about that afternoon's work. A gate in the pasture fence would make a good spot to set up the temporary corral.

"We can open that gate and then run them heifers in," Doug said. "Then kick the heifers back this way." He worried for a moment that the opening would be too tight. "We could brand over here, and just let the panels make an alleyway so they go straight out," Doug said, motioning toward the open pasture east of the fence. "Right?"

"I think so," Leif said, and he pointed that same direction. "Get a bunch of horse trailers on each side so they make a wing. They'll funnel in here."

It was settled, and Doug said, "Let's go eat first."

The high school boys lolled on their backs in the shade beneath the horse trailers. Beau's young kids showed up, toddling around the folding table. Leif filled his plate and headed for the shade of a trailer. He took a seat on the ground next to Beau, who is younger but, like Leif, a leader in the crowd. The two men sat quietly, their plastic forks scraping paper plates. Then they talked about problems with fences.

"What's Todd charge a foot?" Leif asked.

"I don't even know," Beau said. "I honestly couldn't even tell you. For a four-wire, barbed wire? I don't know."

Beau's manner was calm and confident, yet humble. Both men were invested in the place itself, but aware that living so close to the land brought challenges bigger than people could overcome. Leif said, "Had a guy from Glendive do some fencing for me. Name is Ron Wells. God does he put a fence up. It's kind of spendy, but you ain't going to fix it for a while. He puts it up right."

Time turned on the prairie.

"That fence down through some of the coulees has gotten so

old," Beau told Leif. "We've been running some of the yearlings down there. That is not good, with poor fence and yearlings. Wayne Jacobson, I wasn't his favorite guy last summer." Beau said that his young cattle had broken through the old fence into Wayne's field. "They were up in his peas," Beau said. "Thankfully, it was early enough. It didn't hurt nothing."

Wind tugged at the grass.

Like the Jelleseds, Beau's family had many oil wells drilled on their land. Such wells brought money for landowners and jobs that allowed farm kids to stay close to home. But fences no longer kept things contained. The earth had opened, and the prairie above, and the people living on it, were already controlled by all that came to the surface.

There was long silence.

Leif told Beau that he got a call a few days before from someone at Hess, an oil company drilling in the area. "I might have to move my bulls," Leif said. "They had a leak in the pipeline. Fluid got out and got in the live creek down there. They've got this stuff to hold back the water and the oil and stuff to make sure it's not going anywhere in the stream. I said, 'Well, how long did this thing leak?'" Leif said that the Hess workers had told him they didn't know, exactly, but they figured only a barrel of fluid had entered the creek.

"That don't make a lot of sense to me, does it?" Leif told Beau. "They can tell me how much there was, but they can't tell me how long the leak was." He paused. I sat beside Leif and listened as he continued to tell Beau about the single rupture in the vast network of oil, a spill unfelt from a great distance, but not when rooted in the prairie. "I don't know what I'm going to do," Leif said to Beau. "It's a pretty live creek. It gets a lot of water through there."

※

Dig into the soil in Charlson, North Dakota. Go straight down, through clay, silt, and gravel, sandstone and limestone, shale and dolomite. Drill 100 feet down, 1,000 feet. Keep going, layer after layer, churning through rock containing coal and leonardite, ash and gas. Follow the steel drill pipe as it travels through time, penetrating rock that is 10,000 years old, 5 million years old, 65 million years old. The Paleocene period gives way to the Cretaceous, the Cenozoic era to the Mesozoic. Then 10,000 feet down and 365 million years ago, stop at a single layer of shale that is, more or less, 110 feet thick. That layer is called "Bakken."

The Bakken Formation was discovered in the early 1950s, and it first produced oil with a rare strike on a well drilled on the land of farmer Henry O. Bakken north of the Missouri River. Among the many words used to describe the Bakken shale are *carbonaceous* and *fissile*. Carbonaceous, in this case, means crude oil, the remains of small sea creatures and aquatic plants covered and cooked and compressed within the rock during millions of years.

Such a discovery wouldn't have meant much to people prior to the past century. While cultures have known of oil for thousands of years—Babylonians used petroleum as a caulk, a medicine, and in the building of roads—it largely remained a curious substance with relatively little value until the 1850s. At the end of that decade, at a time when whale oil and candles were still the main source of artificial light, prospectors in western Pennsylvania drilled sixty-nine feet deep near a creek and struck a deposit of crude, helping to set in motion an evolution of oil that has since changed the way the world works. A chemist in Philadelphia had figured out how to distill the petroleum into a decent lamp oil, and three years later, in 1862, three million barrels of crude were claimed from Pennsylvania wells. When cars came on the scene decades later, the oil industry continued its acceleration, as crude was converted to gasoline. Oil was discovered in Kuwait in the 1930s and Saudi Arabia in the 1940s, and it became the driver of global economic markets. Now, just over 150 years after the Pennsylvania strike, the typical forty-two-gallon barrel of oil gets refined, on average, into twenty gallons of gasoline, twelve gallons of diesel, and four gallons of jet fuel, with six gallons left for use in the manufacture of chemicals and plastics.

Despite the desire for ever more, back in Henry Bakken's day it was nearly impossible to access the oil that was tightly compressed and widely distributed throughout the rock ten thousand feet beneath the North Dakota prairie. But then at the end of the past millennium, a powerful combination—horizontal drilling and high-volume hydraulic fracturing, or fracking—shattered deeply buried rock in Texas and released its bounty of compressed crude. The industrial system was thirsty, and the profits sure, and the advancement of fracking technology triggered a billion-dollar expansion of oil and gas exploration, with recent booms in Texas and Arkansas, Pennsylvania and Ohio, and elsewhere.

In North Dakota that other Bakken word—"fissile," which means capable of being split—offered promise too. And North Dakota law gives near-complete control to corporations willing to claim oil for financial reward. As one local observed of the order in the state: "mineral rights trump everything." So companies from around the world rushed into western North Dakota and drilled down into that layer of shale 350 million years older than any human species.

By May 2014 a million barrels of oil a day were flowing up to the surface. Dozens of rigs were anchored atop fields in and around central McKenzie County, eager to drill more. The morning after I joined Leif, Doug, and the others at the branding, I shouldered my backpack, turned south at the red wooden barn that Leif's grandfather had built, and walked toward the epicenter of the action.

It felt right to be moving on foot. Walking is the very thing that advanced evolution some two million years ago, when *Homo erectus* straightened all the way up and left, for good, the trees. It had taken four million years of skeletal transformation—a different pace of technological progress—before *Homo erectus* was able to walk farther and faster than all humans who had come before, and for two million more years after that, it was often at the speed of single steps that things advanced, as humans migrated out of Africa. Now many of the seven billion people settled around the planet more often travel at sixty-five miles per hour behind the wheel of a car or five hundred miles per hour suspended in a steel tube in the sky. All the while they sit, muscles still, blood flowing more sluggishly from lack of movement. It takes a physical toll, to be sure, but also psychological. I knew this from the routines of my daily life at home in New Hampshire, where my wife, kids, and I shuttled between school and errands, work and entertainment, in one of two cars. I would drive to a gym for exercise but seldom experienced what it meant to exist in the elements.

The evening before I began my walk on Sand Creek Road, I had strolled around the campground at Tobacco Garden, and not long before twilight I saw two young boys swinging a long stick as though it were a baseball bat. The boys were six or seven years old, so the swinging seemed as much about idle entertainment as practicing any sort of game. One boy would swing the stick for a while, as the other sat on the ground. The boys were suspended in that golden moment between dinner and the call of an adult for bedtime. The next morning, as I took my first steps toward the open prairie, I passed the spot again, and I saw the stick lying in the grass, forgotten. The old branch was five feet long and an inch in diameter. The surface was smooth and had pronounced knots, one at shoulder height, a good notch for a handle. The stick was remarkably light, yet stiff, a rare find in a land with few trees. I decided to carry it with me.

After the long day of branding at the Olson Ranch, I was excited to be moving alone again under my own power, and I held the stick in my right hand, planting it on the dusty road with each step. As the poet Gary Snyder writes, "We learn a place and how to visualize spatial relationships, as children, on foot and with imagination. Place and scale of space must be measured against our bodies and their capabilities. A 'mile' was originally a Roman measure of one thousand paces."

I swung the stick confidently as I paced south from the Jellesed's big red barn. It was Memorial Day weekend, and the prairie roads were empty at that early hour. I passed the low ranch house of the Thompson farm and just beyond that several oil wells. A sign by a gate warned of the presence of hydrogen sulfide, but there was a steady breeze, and the air smelled of sulfur, an indication, I'd been told, that the gas was not present in a dangerous form. After another mile I crested a hill and saw the white steeple of a church far to the south. It seemed adrift in fields covering an area that, millions of years before, had been sunken beneath a shallow sea.

My pack, in which I carried a week's worth of food and gear, hung heavy on my back, and my breath was halting. But I was happy to be working hard. My legs ached but were getting stronger and more steady with each step. I was not as hungry during the day as I could be at home, sitting at my desk. I rationed water, drinking regularly, but only what I needed.

I followed a grid of dirt farm roads toward the church: a mile south, then west, then south again. I made another turn west and passed six wells, these with flares burning off natural gas that came up in pipelines alongside oil. The gas had value as fuel too, but the oil wells had been drilled so fast there was not yet a network of pipelines to carry the gas to market. So it was burned on the spot, as waste. The flares roared as loud as a jet plane taxiing. But several hundred paces farther, that was distant din, and I focused instead on thick grass in a ditch alongside the road. Two pheasants jumped just a few feet to my left, and I jumped too, then called out encouragement as they took flight.

The church loomed larger as I approached, and suddenly I was cutting across a tidy lawn and past straight rows of gravestones, including that of Leif's father. A sign outside announced that Sunday service was being held that week at another prairie church ten miles away. The door was unlocked, though, so I stepped inside the small chapel, then wandered downstairs to a kitchen where a hand-painted wooden sign offered a promise: "Drink Lutheran coffee. It's heavenly!" The pot was unplugged. I went back outside, sat on the front steps, and pulled an apple from my pack.

Across the dirt road, just a few hundred feet to the south and west, more well pumps churned, drawing up oil from the Bakken shale ten thousand feet below. The wells operated with automated precision, storing oil in tanks until trucks came to haul it away. At odd intervals as I ate my apple, stack pipes roared and flames of more flares leaped in the sunlight. The little church stood silent, as though overwhelmed by the relentless labor of the wells, especially on a Sunday morning.

I sat between the stoic steeple and raging flares, worlds colliding beneath blue sky swept with sweet spring breeze. More than sixty wells framed an eight-mile stretch of the road alongside the church. Locals who farmed and ranched nearby called the gauntlet the "Midnight Run," so redefined it was by the industry of oil. I took a last bite from my apple and hoisted my pack onto already aching shoulders.

Less than a mile east of the Lutheran steeple, two white pickup trucks stood near the center of a well pad. The square pad, like most of the others, was more than two acres in size and had an even surface of crushed red rock, called scoria. At the center stood a pump and several tanks, and there two men—one older and large, the other younger and lean—huddled around an idle rod, their hard hats nodding toward their work.

When I first saw the men, I hesitated. I did not know the rules of the oil field, particularly as they might apply to a random visitor such as me. My arrival was certainly unexpected, but I meant no harm. Just as the ranchers at the branding had not at first known what to make of me, I was, standing by the well pad, hesitant about my place. The gate to the pad had a "No Trespassing" sign, and another warned of the possible presence of H$_2$S. I stood in the road for two or three minutes, leaning my weight and that of my pack onto the walking stick. We hovered there, the two men working and I, suspended in odd proximity, a distance of a hundred feet between us but bounded all around by treeless prairie. Then I walked through the gate.

The younger man glanced my way, and I waved my hand, as if to ask permission. He turned back to his work, and I took that as a yes. When I arrived at the pump, there was a moment of silent greeting as I slid my pack to the ground and then sat on the dirt. I said hello and so did they, and for a few minutes it was just like that, two men working and another who had walked off the horizon and taken a seat.

Each man wore a hydrogen sulfide monitor clipped to his shirt, and their job that day was simple: servicing the polished rod that connects the pump to underground parts of the well. They had stopped the pump, releasing pressure, and the younger of the two men was doing the most essential of the work. His hands were covered in blue rubber gloves, and those were covered in thick globs of grease. The older man handed him heavy metal rings to slide into a casing.

They told me that they worked together as a maintenance team, seven days on, seven days off, for ConocoPhillips. That afternoon, already warmed to more than eighty degrees, was a good one for the job. They stood out there fixing casings in February too, when temperatures could drop to thirty-five degrees below zero, and brittle winds blew. On such afternoons they kept the pickup trucks idling, hopping out to work five or ten minutes at a time, until hands grew numb, then taking shelter again in the heated cab. "Someone has to keep these things running," the younger man said.

His name is Ryan Bak. He was twenty-seven, from Billings, Montana, and had studied petroleum engineering in college but took this job because he could balance it with life back home. He said he was staying away from bars in Watford City and commuting to Montana every other week to see his parents and sister, who was confined to a wheel chair. "I won't have them around forever," Ryan said. "And I'm not going to leave that to chase the dollar."

Ryan strolled to the bed of his truck, dipping his hands in a container for more grease. The older man, whose belly protruded so that he had to extend his arm to reach beyond it, was in more of a settling mood. At fifty-six, Harold Wilson had worked for years in a mine in the thickly forested canyons near Challis, Idaho, helping to harvest molybdenum, an element that can withstand extreme heat and is used in everything from light bulbs to computers to space vehicles. But work there had slowed, so he left his little house in the Rockies and, like so many other people, towed a camper trailer to North Dakota to call home. He had recently bought ten acres down near Grassy Butte, and he planned to build. "Living in that fifth-wheeler's not my idea of fun," Harold said.

Ryan talked about the fact that far from the prairie—he mentioned Wall Street as one example—people cheer for more oil to come out of this well and others. If the well has a problem, those people far away don't care if a guy like him has to come out to fix it in bitter winter or blazing summer. But Ryan also noted that the people who profit only can do so because of everyone else counting on wells like this not to stop. "Who's willing to give up their big-screen TVs and Xboxes and all that other stuff?" Ryan said. "The hard, ugly truth is you need this. It just has to be done."

He finished packing the casing and turned the well back on. The pressure climbed as the pump rose and fell again. Ryan stopped next to the tailgate of his pickup, removed his gloves, and wiped his hands. He rattled off a list of products that contain petroleum: shoes, bottles, cans, electronics, gasoline, jet fuel. "Oil is in everything you do and use," he told me. "If it's not in everything, it transported it to you."

He praised the quality of the crude that comes up from the Bakken, so pure it could almost be poured directly into a gas tank. "Energy can't be created or destroyed," he said. "It can only be stored." He marveled at the changes that came from burning fossil fuels during the past century, especially during the fifty-year span between the first flight and a human landing on the moon. As Ryan talked, he and Harold prepared to climb back into their trucks and hustle off to service another well. But first Ryan looked at the early edge of this century, and he was optimistic that the oil would carry humans to new and exciting places. He marveled in his question about the years ahead: "Where will we be in fifty more?"

I said good-bye to Ryan and Howard and, walking east, soon came to a wide, shallow stream and saw several large birds along the shore. I did not know whether the birds were storks or pelicans, both of which live in North Dakota. Just beyond the pond rose a work-over rig, a contraption brought in to freshen up wells that have already pulled a lot of oil from underground. The presence of the pond, and this second well, jolted me from that intimate encounter at a single oil pump back to the larger reality: the harvest was happening everywhere, and I knew that for days my route would continue to weave through the oil field.

For that I carried a map, a single page torn from a road atlas made for oil-field truckers. Prairie roads, originally built to serve farming and ranching, were set in a loose grid, with east-west roads intersecting north-south roads every mile or so. The oil crews were building on the network that came before. So the colored dots on my map that showed the location of drilling rigs, and hundreds of smaller gray dots indicating producing wells, were set along east-west roads two miles apart. The wells were spaced on the surface because once a drilling pipe reached two miles underground, it turned horizontally and ran another two miles north or south through the thin layer of Bakken shale. The idea of the companies was to make sure the underground well pipes did not collide but still came close enough to reach all the oil, everywhere.

I had five miles to cover before I would arrive at an alfalfa field where Leif had said I could camp for the night. I turned south on County Road 1806, its two paved lanes splitting more flat fields, and I saw up ahead a steady stream of tankers, gravel trucks, and one-ton pickups turning east and west. I pulled out my map. One red dot indicated that to the west a drilling rig was working for Burlington Resources. A green dot noted that a rig to the east was drilling for Hess. Lines of smaller, shaded dots confirmed there were already eighteen working wells in that short stretch. I did not have the gumption to engage with more machines.

A mile farther south Highway 1806 intersected North Dakota Highway 23, a high-speed road that carried traffic between Watford City and the Fort Berthold Indian Reservation. There I found an unexpected oasis: a worn yellow camper trailer converted to a food truck. I had in my pack freeze-dried meals, and carrying them and everything else had become an exercise in a kind of self-sufficiency. But twenty-three miles on foot had sharpened my sense of basic survival and the need for calories that would carry me farther. I felt something no doubt shared with those early *Homo erectus* and many more who followed: take food when you can get it.

I dug out my wallet, then looked through a small window at a man who had lost his job in Seattle and now tended a hot griddle. I ordered a burrito to eat then and a hamburger to carry with me to camp. A large dog was chained at one end of the trailer, but he seemed little bothered by me. The man behind the window handed me my food, and I walked around the other end of the camper to the shade of a lone tree. I dropped my pack and hunched over my meal. I was sweaty and tired but knew I needed to eat. After hours of moving on small snacks through the hot day, my body craved more calories. I sat with my back against the tree trunk, and soon the burrito was gone.

Across a dirt lot, a bare-backed man fired a nail gun as he built framing inside what looked to be a long-abandoned building. He bantered loudly with two other men in a riotous conversation, shouts and cackles, insults and laughter. One man stepped outside to fetch more lumber, and I decided then to keep my head down. Before setting out on my walk I had been warned about rough characters in the oil field. I noticed that another one of the men, his tattooed arms swinging loosely, was striding across the lot directly toward me. I didn't have time to stand up, so I opted for submission, lifting my eyes toward him as he stepped in close, blocking out the sun. "Take it," he said, holding a brown bag at arm's length.

I rose to my feet and looked inside: three sweating bottles of cold water and a six-pack of Busch beer. I laughed and explained that I didn't need it. He insisted. I offered to pay.

"No way," he said. "We looked out here and saw you sitting under this tree on such a hot day, and we just said, 'that ain't right.'"

The man went back to work, and I sat again. After a few minutes I saw a tick climbing up my bare shin. The tick moved slowly toward the hem of my shorts. I flicked it off. It came back, climbing with purpose from my boot onto my leg. I flicked it off again, and it landed several feet away. The tick returned to my leg, determined to dig in and feast on blood. I picked it off and put it on my left thumbnail. The tick's back was deep brown with a reddish rim. Its eight legs were short and thin. It needed my blood for a survival that threatened me. I turned my right thumb upside down above my left, then pressed my thumbnails together. I held steady pressure until the tick popped.

※

When I had first begun to plan my walk at home that spring, I knew from earlier jobs as a news reporter—covering courts and companies, politics and social issues—that it often pays to work through established channels. So I sent emails and made phone calls to press officers at oil companies drilling in McKenzie County. I explained over the phone to a field-operations manager for Petro-Hunt that I would be walking through the area and that I would like to tour one of the company's oil-drilling sites to see how things worked. He listened patiently and then said, "We are going to choose not to participate in this opportunity." I left five voice mails for the chief media-relations officer for Hess Oil Company at his office number in Houston, explaining in one message that he, a former reporter for the *Houston Chronicle*, and I, a former reporter for the *Boston Globe*, would understand each other when we talked about access. He never returned a call.

But I also had learned during a decade of reporting about life in remote parts of the world—whether in sweltering valleys of Ethiopia or cool forests of Maine—that the most enlightening stories often emerge at ground level. I had sat with date farmers on the banks of the Nile, reindeer herders on the frozen tundra of the Russian far north, and migrant farmworkers in California. In such moments I learned more about life in those places than I ever could have in the halls of power.

So by the time I walked south from Highway 23 and the food truck that Sunday evening, I was glad to be moving freely, without any prearranged plans. Eager to make camp for the night, I strode south on a dirt road toward the alfalfa pasture, with cold beer and warm burger swinging in a bag in one hand, my walking stick in the other. Moving on foot, I was a curiosity, at least, and a person in need of help, at most. My backpack signaled to strangers that I was far from home and close to the edge. I explained to anyone I met that my goal was to experience and to write: to tell stories of what I found. Because of my vulnerability, I think, most seemed more welcoming. A friend who was following my journey from afar sent me a note via Twitter: "You've entered the world that nobody arrives to by accident, which makes you part of the team by default."

Then I saw up ahead in the dusty twilight, coming my way on the long, straight road, what I thought must be—but could it be? —someone, like me, walking. The person was a quarter mile away, only a shadow on the horizon. I saw in the distance, on the right side of the road, a ranch building. But ranchers no longer move on foot, choosing instead pickup trucks, ATVs, or sometimes a horse. On the left side of the road, an oil-drilling rig towered more than 150 feet high. As we drew closer, the man and I called out greetings, and I stopped and leaned my weight onto the walking stick. The man's Carhartt T-shirt and Halliburton Drill Bits and Services cap confirmed my suspicion: Kyle Gray, age fifty-three, had just finished a shift on the rig and gone wandering for fresh air.

A shine in Kyle's eyes overcame weathered cheeks, and we stood fifteen minutes or more, talking as neighbors on a porch. Kyle knew the ranching life, as he'd grown up herding cattle in southern Utah. But he told me that for the past thirty-one years he'd been working on rigs in oil and gas fields. He spent a good bit of that time in the natural-gas boom around Pinedale, Wyoming. He'd once taken a picture there that showed twenty-eight drilling rigs in the frame of the photo. "We could throw rocks at each other," Kyle said. He'd first come to North Dakota in 2006, when oil companies were beginning to explore the Bakken. He returned for a longer stint in 2010. Watford City—a place I would see after walking thirty more miles—had been nearly a ghost town the first time Kyle visited. By the second it was packed with worker camps. "Watford City just got hammered," Kyle told me.

I wrote down Kyle's cell phone number and email to keep in touch when my walk was done. Just before I moved on I asked Kyle what, exactly, he did on the rig. He told me he was an independent contractor working as an on-site supervisor for the oil company, in this case Hess, to dig new wells, in this case the HA-Thompson 152-95-2017H-3, -4, -5, and -6. "I'm the man in charge out there," Kyle said. "They hire me to come out here and drill the wells no matter how it has to get done."

I was worried about making camp before nightfall, so I wished Kyle a good walk and continued south. It was already nine o'clock, and I needed to cover two more miles. The alfalfa field lay near the edge of the Blue Buttes, a pocket of peaks and canyons. The shadowed contours of the closing buttes messed with my sense of security. That worry scuttled my sense of progress, and I wondered if I had passed the alfalfa field. At one point I lifted my pack over a barbed-wire fence and climbed into a pasture. I did not know what alfalfa looks like, but I could tell that the pasture was clearly prairie grass for grazing. I walked the fence line to the top of a hill. Below I saw another field, and there a shallow pond, which Leif had also mentioned. Ankle-high tufts of fuzz-tipped leaves trailed up a slope to the west. I climbed the fence again and hurried to pitch my tent.

A natural-gas flare on a hilltop to the west fluttered gold. The flame bolted in the wind, as shaken as a just-branded calf in an unexpected moment of evolution. The green field turned toward black, creating an odd mix of nature and industry. Everything seemed to merge as darkness came complete. I could hear the rumble of distant trucks on the road I'd walked but sensed the seclusion of the field, surrounded for a half mile in every direction by the fluff of alfalfa. As I lay on the ground, I realized that for the next several hours I would be most at risk, as mountain lions and coyotes lived in the buttes and could pass my way.

So exposed in nature, I was beginning to find a more personal understanding of a broader threat: society's dilemma in the face of climate change. The burning of fossil fuels around the world emits so much carbon that temperatures are rising. Flooding, drought, and other extreme conditions in otherwise less volatile parts of the planet are accelerating. Scientists warn that if rising tempera-tures are not halted soon—at a total of 3.6 degrees Fahrenheit by the end of this century—a tipping point could be crossed for the world as we know it. But the problem, when considered amid the distractions of daily life, dislocated from the source, can seem too vast. How does any one person, hurtling from day to day in a world run more or less on unseen fossil fuels, even begin to con-nect to or care about incremental damage to the planet?

Sleeping at ground level, I could sense the danger that came with opening up the earth and uncorking carbon to support lives on the surface. As an animal, alone, I was vulnerable, needing na-ture more than trying to overcome it. I slept fitfully through the night, waking often to see the orange wash of the oil-well flare dancing on the thin tent walls.

The next morning daylight brought different exposure, so I packed quickly and moved closer to the road, where I set my things behind a water tank next to the pond. I huddled behind the tank —out of sight of passing trucks—and boiled water for breakfast. Even though I had come so close to the source during my first days of walking, the most essential structures, the drilling rigs, still loomed as isolated islands. I could hear the drone of the drilling as I passed each rig, but I could not feel its force. I wanted to see the ancient earth from which the oil came.

As my tea water rose to a boil, I flipped through my notebook, then dialed Kyle's cell phone number. I told him I could walk back to the HA-Thompson drilling rig in an hour. I asked if I could climb the rig to see how things worked. I was anxious as I spoke, wondering if this might be my only chance. Kyle did not pause for even a second but barked a reply with the unbridled certainty of his Utah youth: "Absolutely!" He told me that when I arrived at the well pad, I should walk to a gray trailer at the back of the lot, where he kept an office. "Anyone gives you any shit," Kyle said, "just tell them you're coming to see me."

An hour later I trudged up the long dirt road to the entrance of the HA-Thompson operations. The well site appeared as a moonscape, the soil scraped back and replaced with crushed red rock. Dozens of trailers lined the right side of the pad. Pickup trucks shuttled in and out, as did a tractor trailer. The driver stared down at me, approaching with my pack and stick, as though I were an alien species with suspect powers. I strode toward the towering rig, then angled to a trailer dwarfed beneath it.

Kyle answered my knock at the door, and soon I was sitting in a soft seat, sipping a cup of cool water, as Kyle talked about depth and drilling and dates of completion. I told him what I really wanted to do was get up on the rig floor, fifty feet above the ground, where the work was done. Kyle leaned back, stretching his arm behind him, then rocked forward and tossed a hard hat into my hands.

Outside we climbed flights of steel stairs and stopped in a small room, where Lance Frick was settled into a swiveling bucket seat. Lance is by some measures huge, born on a farm on the Great Plains and grown big enough to play offensive lineman in college in Montana. All his strength was steering a joystick, as his eyes glanced out the window onto the open deck at a pipe slowly descending. The drill bit at the head of the pipe was nearly four miles away: two miles straight down and then, after a big turn that began at 10,150 feet, the *kickoff point* they called it, the pipe ran horizontally two miles through the Three Forks shale, just below the Bakken.

The drill bit leading the way was, at that moment, 10,684 feet underground, and it had run off its course. So the drill bit was "sliding," which is to say, turning back in the right direction, and the going was slower than when the pipe is driving straight. That is called "rotating," and nearly everyone on the rig liked it better when the pipe was rotating, because the well was drilled faster. If this particular pipe didn't get turned straight soon, it might drive up and into the zone of another well from this pad, which had already been drilled sideways through the Bakken. Lance sat in his pilot's perch, watching a bank of computer screens overhead showing bit depth, hook load, differential pressure, and something called flow. He took a moment to reflect on what he was doing. "It baffles my fucking mind is what it does," Lance said.

Dislodging deep rock in a precise and violent way takes a lot of coordination, and of all the computer screens covered in dust and grit hanging in that cabin of control none marked the goal of Lance's work as clearly as the one with a bull's-eye graphic of concentric circles. Those circles showed the direction of the steerable drill bit deep underground, and, across the cabin from Lance, James Drake kept a close eye on that target. James wore a bushy beard beneath his black hard hat, which was emblazoned with a longhorn steer. James is a horizontal-drilling specialist who works for Baker Hughes, and he is known as an MWD hand, which stands for Measurement While Drilling but often gets turned into Movie Watching Dude. Killing time is what James and other horizontal drillers often have to do when a pipe is on course, rotating toward its target—but not when the pipe is sliding, so James checked seismic surveys.

"Give me about 290," he shouted to Lance, who answered the command with a swivel of the joystick.

James was trying to get the drill bit swung around from 135 degrees. "I was at 219 on the last survey," he told Lance. "We're almost there."

Those last 71 degrees of sliding would take some time, so most of the men working that shift—floor hands named DJ and Chris; another guy who goes by Mississippi; the motor hand, Rex; and the derrick hand, Richard, known as Banger out on the rig—were busy elsewhere in the network of steel stairs and platforms that cascaded down to ground level. There, among rows of dozens of trailers, Kyle could keep track of benchmarks jotted on a dry-erase board on the wall of the supervisor's office. The drilling of this well had begun eight days before, and if all went as planned in those final hours, its horizontal reach would be complete the next day.

The rig was one of more than a hundred drilling in the Bakken at that moment, and each was in high demand. The big rig would be disassembled and hauled, a few days later, north across Highway 23 and past the yellow food truck, then east on that road with the string of wells, where it would be put back together to drill all over again. Back on the HA-Thompson well pad, the water tanks would roll in, and the hydraulic fracturing would begin: millions of gallons of water mixed with potent chemicals would be pumped at pressure down into the Three Forks, where the collision would splinter the shale and release the oil.

The technology was effective, if not subtle, and celebrated by engineers and executives eager to unlock oil and gas deposits in places not before possible. During the decade and more before my North Dakota walk, hydraulic fracturing helped to increase U.S. production of both oil and gas by more than 50 percent. The oil, more than eight million barrels of it a day, was flowing up primarily from the Eagle Fork formation of south Texas, the Permian Basin in west Texas, and the Bakken and Three Forks beneath western North Dakota and eastern Montana. Gas was harvested from those places but also from the Fayetteville shale in Arkansas and the Marcellus and Utica shale formations beneath Pennsylvania, Ohio, and other parts of the Appalachian Basin.

Such fracking is rife with dangers, primarily from pumping so many chemicals through layers of earth that contain water, but also because of the swaths of landscape that it industrializes. Opponents of the process were outmatched by the velocity of the system. Their cries of warning from afar echoed through town meetings and the halls of Congress, while workers on the rigs dug fast, increasing the production of fossil fuel.

On the rig drilling the HA-Thompson wells, as Memorial Day turned from morning to afternoon, there were several hundred more feet forward for the pipe to travel underground. So as Lance and James drove the course of the pipe in the rig cabin above, Banger walked along a balcony halfway up onto the rig and kept an eye on the shaker, a vibrating conveyor belt that delivers the cuttings, as the earth brought up from below is called. Water was pumped down into the hole to ease the drilling, and the cuttings arrived at the surface as a combination of liquid and solid that seemed as if it could have been pulled from the ocean floor. As the shaker rumbled, a thin flow of the watery gray sludge slopped into a chute. I reached out and wiped my fingers through the pre-historic grit.

For millions of years this once-solid rock had held the carbon deep underground. Within a few moments it had been shattered for fleeting use. The cold and heavy mass—slopping onto the shaker in a steady flow as the drill dug deeper—would be pumped to a tank and stored as waste.

Up on the rig floor it was soon time to attach another thirty-foot section of pipe, and Lance put out the call to the workers scattered below. Connecting a new section of pipe is one of the few jobs that requires rig workers to put themselves in harm's way, fitting hydraulic clamps, setting cables onto pipe sections, and standing clear as the iron swings overhead. DJ and Chris climbed the steel flights and strode onto the rig floor, hard hats set tight atop their heads, fire-retardant shirts and pants hanging loosely on their muscled bodies. They hoisted clamps and pulleys with a certain urgency, like soldiers at the front, scrambling to protect the world beyond them, as though all would collapse if they did not keep going.

As I watched, it seemed the turning of the pipe deeper into the ground would never end, and after a few minutes more I told Kyle that it was time for me to move on. I did not want the dynamics of the drilling to begin to seem too normal. A deeper reckoning of its toll would come with distance.

Four hours later I was four miles southeast, in another situation completely: no towering structures climbing off the prairie, just another landowner and his farm machines scrambling low against the soil in the season of planting. Don Nelson, whose grandparents, like Leif's, also settled in the prairie a century ago, had spent that Memorial Day shuttling seed out to his fields. He was about to break for lunch when I wandered into his farmyard, and he invited me to join him. Soon he was cooking steak and warming leftover mac and cheese, and we sat together at the counter in his kitchen.

Don is worn and weathered from working the land, but we spoke more about oil than wheat during that meal. There had been some oil drilling in McKenzie County in the 1950s and 1980s, but that was limited and did not redefine life on the surface in any significant way. The Nelsons hadn't profited during earlier decades, but the depth of the Bakken boom meant that Don had several new wells on his land. He collected good royalties, but those did not come easily. The next week Don and some cousins would travel to a state hearing to fight to keep payments negotiated with Continental Oil Company for wells on family land near the Missouri River. Don knew that any financial gain from drilling for oil came with deeper risks.

"We don't know what's going to happen here," he said between bites of steak. Water is everything in the prairie, whether flowing on the surface or deep underground. "They're punching holes through all that stuff, and it's a man and a machine, and sooner or later something's going to fail."

Don and I went back up to his sheds. Spring had come late that year, and he was hustling to make up for lost time, now that the soil was warm and dry enough for planting. He backed his grain trailer beneath a small silo and started to fill the trailer with seed.

His plan was to shuttle it out to a distant field that afternoon so his son-in-law would have plenty to keep planting.

My first four days walking the prairie had been graced with clear skies, so I did not know the nuances of prairie clouds. But even I, sitting nearby as Don climbed up and down the trailer making sure it was filling fast and even, could tell that the black bank of clouds looming so suddenly in the western sky was not a gentle thing. There was blue sky north and south, but turmoil at the edges of the giant V. Don saw it too, and he backed the trailer under the cover of a shed, then called his wife, Rena. The three of us hustled to get pickup trucks into more sheds, preparing for a battering round of hail.

After descending into a notch of a valley in which Don had built his timber home, we settled into folding chairs in the open doorway of the garage. Rena pulled up a cooler full of cans of cold beer, and we stared at the wooded gully as the heavy rain began. Rena worked most days in the office of the police department in Watford City, more than thirty miles west, and she started dialing the phone. Don's mother, who lives in Watford City, reported that golf ball–sized hail was already falling there. She mentioned to Rena rumors of a tornado. Twisters are rare in western North Dakota, especially that early in the season. Then Rena called her son-in-law, who was still running the tractor out in a field, and he said he could see lightning cracking in the west and bright sky in the east. A few minutes later Rena's phone rang, and it turned out the tornado talk was true.

We moved inside to watch the local news: the hailstorm had more power to the west, and a tornado had formed south of Watford City. There it had swept across a low hill, jumped a two-lane paved road, and touched down in an RV park, a makeshift settlement of workers from around the United States who had come in search of economic salvation. That Memorial Day evening Kyle was continuing to lead the work on the HA-Thompson wells, as were others on drilling rigs across the prairie. But many of the crews supporting their efforts—truck drivers and welders and roustabouts—were enjoying the holiday at home in their trailers. The tornado grabbed eight of those RVs, many of which had people inside, and pulled them into the air. It turned the RVs upside down, ripping off doors, windows, and walls. The agitated currents of air dismantled the homes piece by piece, then discarded everything and everyone back to the hard earth.

＊

Thomas Merton, the Trappist monk and spiritual writer, said of a journey to the source: "The geographical pilgrimage is the symbolic acting out of an inner journey. The inner journey is the interpolation of the meanings and signs of the outer pilgrimage." As I took my first steps west from Don's house the morning after the tornado, walking in a soft rain among the buttes, I was, geographically, halfway to Watford City. With each mile I had walked toward an emotional understanding that what was happening in this place supported my insulated life at home in New Hampshire, and I could not avoid the conflict of my own consumption. My interior journey—that interpolation of the journey's meanings and signs—focused increasingly on the unnatural toll of the oil boom.

My route from Don's followed the mile-by-mile grid of dirt farm roads, turning west, then south, then west again, rounding an arc that took me back through the area where I had joined Leif Jellesed for the branding of Doug Olson's calves several days before. At a point where the road made a ninety-degree turn to the south, a gravel lane continued straight into a pasture and over a rise. A mark on my map noted that this was the site of the Jellesed SWD, which stands for "salt water disposal." This particular disposal was set on more of Leif's land. I did not break stride, so unremarkable was the access road. But day in, day out tanker trucks were turning up that road to deliver thousands of gallons of wastewater from drilling rigs, fracking operations, and producing oil wells. The naturally occurring brine also contains toxic fracking chemicals, oil, and gases. From the Jellesed and other SWDs, the waste water makes a one-way trip five thousand feet underground, halfway to the Bakken shale. Geologists guess that a sandy layer deep underground will contain the water more or less forever. The plan comes with simpler risks at the surface. One month after I walked past the Jellesed SWD, for example, a pipe at the site broke and twelve thousand gallons of wastewater spilled onto Leif's land.

After a few miles more of walking, I saw that on the left side of the road thousands of square feet of soil had been dug up. Tumbled stacks of pipe crowded the pit, and the long arms of two cranes swung above it all, hoisting single pieces and lowering them into trenches. Ray Gilstad, who had done much of the branding, owned the land, which would soon become a key pipeline hub for shipping Bakken oil to distant markets.

The oil field had sprung to life so fast that the system had not been able to keep pace with the arrival of crude to the surface. For the first several years most of the oil had been trucked in tractor trailers shuttling twenty-four hours a day. Those trucks delivered the oil to tankers on railcars that rolled off into the United States. Those were so precarious, exploding in faraway places such as Lynchburg, Virginia, that they'd become known as "Bomb Trains."

Pipelines were touted as the answer, as underground grids would deliver the oil—unseen and out of the way—to distant refineries. But those also had an impact. Two years after my walk Standing Rock Sioux and their supporters would protest construction of the Dakota Access Pipeline beneath the Missouri River, a few hundred miles southeast of the Bakken. The pipeline, industry leaders would say, was the best way to get the oil to customers. But opponents knew it threatened the water supply of already struggling Native American communities. Government forces would support the movement of oil, and eventually the Dakota Access Pipeline was fed by a series of hubs, such as that under construction on Ray's land, where the harvest of oil, gone from boom to big business, was getting hard wired into the earth.

I stood and stared at the cranes slowly assembling the piles of pipe, much of it still just a jumble of parts waiting to be buried. A blister on the ball of my left foot was pulsing. I took a few tentative steps toward the edge of the pipeline pit but did not have the energy in that moment for any kind of encounter. I turned to continue down the road, then I braced as a man in a hard hat and orange safety vest walked quickly in my direction. He stopped close, leaned in, and asked if I was out for a hike. He did not wait for an answer. He told me he was an electrician from Denver. He had arrived in the oil field just three days before. He stared past me at the tough and humbled landscape and delivered his verdict: "They sure are uglying this place up in a hurry."

That evening I took refuge behind another prairie church, its white steeple also standing starkly above a patch of green lawn surrounded by brown fields. Frogs in a streambed sang. Buntings darted between stout gravestones carved with names of settlers from a century before: Olson, Johnsrud, Rolfsrud, Larson, Hansen. I staked my tent and crawled in sometime before dusk, then waited with closing eyes as wind rose.

By midnight, with darkness again complete, a drilling rig two miles north sparkled at its edges. In the center another four-inch pipe continued its journey underground, pounding down one thick mile and another. Somewhere on that suspended city, workers scrambled, calling out numbers, targets for the drill bit far below. But out on the open prairie their voices could not compete with the beat of machines, and in my tent there arrived only a dull drone, the relentless grunt of industrial conviction.

I lay on my back, exhausted but alert, bracing against the drone of the drill. Then came a leap of light. I rolled over to face the south, and there, on the opposite horizon, the bulbous flame of a single natural-gas flare burned. Outside of my tent birds nestled close to the ground, quiet against the unnatural rhythms. The steeple was awash in the orange of the distant flame, and the gravestones shimmered above a burning lawn.

Morning brought steady, cold rain. I packed my tent quickly and moved my things into the hallway of the church. No one was there, but the door was unlocked. I stepped back outside and boiled water for tea, then ducked indoors and sat on a pew in the small sanctuary. A clock on the wall quoted Psalm 37:7: "Be still before the Lord and wait patiently for him." Thunder rumbled, hurried by gusts that moved clouds west, toward Watford City. I hoped to walk at least ten miles in that direction and ten more the day after that—a final solitary push across the open prairie before arriving amid the crowds of Watford City. Inside the sanctuary a fly buzzed against a window pane, and faces in photos stared down from the walls.

My walking during the past days had sharpened my awareness of time and space. I could judge the size of a truck approaching behind me by the rumble it made on the road. I could gauge with a glance how long it would take me to traverse a sweep of the prairie from one rise to the next, step-by-step. I could hear the flit of bird wings above the grass. I could smell the bite of grease smeared against pipes. And as I sat alone within the four walls of the church, sheltered against the movement of the day, but only briefly, I could sense the acceleration of industry overwhelming nature: the appetite of humans outpacing earth's cadence.

When the church was built a century ago, it loomed as an outpost of progress, lost at the edge in a country then home to ninety-two million Americans. Now it floated like flotsam in the oil field, which fueled just a bit of the movement of more than three hundred million people clamoring in cities far away. My grandparents had come of age in a world with few automobiles, and my children could board a plane and traverse an ocean in a few hours. How could people keep fueling such forward movement?

I was at peace sitting still. Yet I knew, as I sat in the pew, my legs and shoulders aching against the effort of earlier days, that the afternoon's walk toward Watford City would carry me closer to the crush of people come for the oil boom. After an hour of solitude, I packed my notebook and checked for full water bottles. Then, with clouds still heavy overhead, I picked up my backpack, closed the church door behind me, and strode west on wet dirt.

I would cover new ground on the last twenty miles to the county seat, but first, just down the road from the church, I passed the pasture where Annette Olson had set up the lunch of sloppy joes and baked beans during the branding. That pasture was empty, but a mile beyond, in a pasture to the south, I saw the calves. They were idling in the field of greening grass, some standing near mothers, as they'd been reunited, others resting on folded legs close to the soil. From a distance they looked no different for their branding. Did they know?

A pickup truck approached, and I saw that Doug Olson was driving. He slowed to a stop, and we talked through the open window. He confirmed that, in less than a mile, the road I was following would turn to a double-track path that would make the most direct and solitary route west toward Watford City. We said goodbye, and then I was lost as much as a person can be, out on the long-gone sea, my feet moving in a single rut of the double track, the surface more chunky and unsure than on roads. I saw three rigs rising like skyscrapers across the southern horizon, drilling for Burlington Resources on the Haymaker, Shafermaker, and Bullrush well pads. Songbirds darted shoulder high, and I talked to myself about the steps I was taking. To the eye the prairie all around was flat, but to the legs there was constant up and down. I stepped across puddles in a shallow gully and climbed another rise. It was a small hill, but there were so many.

After two hours of steady walking, I stopped at a cattle guard, where the double track turned back to road. I dropped my pack, slumped against a fence post, and ate beef jerky. I was sad for the walk to be nearing its end, despite how hard it had been to carry my backpack miles each day. That effort had brought an intimate connection to the place and its people, and I wondered if that might be overwhelmed once I reached Watford City. But I was also invigorated by the idea of entering so crowded a place. I soon rose from my spot of rest and headed west again. The road traced a high plateau, and to the southwest I saw the sprawl of Watford City in muted sunlight.

The next morning, after camping between another prairie church and cemetery, I walked straight south on the final miles into town. Past a just-fracked well, traffic increased: semis hauling oil and gravel and water, one-ton trucks with welding equipment and concrete, pickups loaded with tools. The machines rumbled by in clouds of dust: Crestwood Crude Transportation, Jericho Services, Baker Hughes, Continental Resources, Franz Construction, and J&L Services. North Dakota highways had become the deadliest in the United States, as hurried oil-field drivers were pushing to get their jobs done. I walked along the left edge of the dirt road, better to see the oncoming trucks. But the lane was barely wide enough for two rigs passing, and soon I heard the growl of a tractor trailer slowing behind me. I pivoted cautiously beneath the weight of my pack, as the truck rolled by at no more than five miles per hour. The driver's window descended, and a bottle of water flew out, landing at my feet. I stopped to pick it up, and before the window rose again, I saw a lone hand extended to offer a thumbs-up.

As I arrived within a few miles of Watford City, the truck traffic was so steady and so hurried that I found myself bracing with each passing vehicle. For the first time I doubted that I should be moving on foot. I staggered often from the road into high grass for safety, not unlike a startled pheasant taking flight.

At an intersection with a stoplight on the east edge of town, oil-field traffic stretched a mile east and south, and I waited my turn to jog across paved lanes to a sidewalk that led west toward Main Street. I stopped for a moment to stare at the mass unmoving, and a man in a sedan rolled down his window, curious about me and my backpack. I asked if he was a local. He said he was from North Dakota, but not Watford City. He laughed and said, "Watch out for the crazies."

When the North Dakota oil boom had become big business, the industry needed everything from pipeliners to bartenders to support it. Thousands of individual Americans—connected to no enterprise other than expectations—packed up their lives hobbled by a collapsing economy elsewhere and struck a claim of their own in and around Watford City. The seat of McKenzie County had been home to fewer than 1,400 people in 2005, but some 20,000 were huddling in and around the place when I arrived. Most of the newcomers were living in trailer parks, paying $650 or more a month to tether their temporary homes to the ground with extension cords and garden hoses.

It was early afternoon, and I walked a few hundred yards more and turned into the Watford City Tourist Park. The park had been built during earlier isolation as a waypoint for summer travelers driving across the United States. Stout trees shaded a green lawn, and a brick building had showers. More than a dozen camper trailers stood side by side in the full parking lot. Wooden posts marked three tent sites, each unclaimed. I lowered my pack at one and sat at a picnic table next to the banks of the brown and slow-moving Cherry Creek.

The door of a nearby trailer opened, and a woman with sandy hair pulled into a ponytail called out a welcome, as though I had just arrived home for the holidays. She told me her name was Louise Tanner, and she and her husband, Richard, had come to Watford City from Colorado. Richard had earned a degree in electrical engineering in his sixties, but he could not refuse the money that he would earn operating heavy machinery for the construction of a highway artery around Watford City. The couple also hired on as hosts of the city tourist park, and Louise took the lead registering new residents and keeping the bathrooms beyond clean.

She told me that when she and Richard had arrived three weeks before, there was daytime drug dealing and prostitution in the parking lot. Louise called in the police. One afternoon she strode up to a group of brash young men and chastised them for misbehaving. "Everybody moved out," Louise said. "Within twenty-four hours they were all gone, and there was a peace here you could not believe."

Louise knelt in her camper each day to pray for three things for the park: safety, peace, and provision. New residents who had arrived during the past three weeks included the director of McKenzie County Emergency Services and her husband, a former sheriff of neighboring Dunn County who was spending his days hauling oil in a tractor trailer; a young couple from Oregon and their three children, who had parked their camper for a few nights while awaiting a move to the city of Minot, where jobs abounded but prices were not as high; and a family who lost their RV in the tornado but were able to buy another worn camper after taking more money against their house back in Michigan. During my days in Watford City, Louise would be a sort of protector for me too. She offered that first afternoon to turn on the hot water in the public showers for longer than eight quarters could keep it running.

I was walking from the showers back toward my tent when I saw two men sitting at a picnic table. The older man, middle-aged, was heavy and unshaven, and he looked as though he had a tough time keeping himself together. The younger man, maybe twenty years old, was harder and more alert, but also looked more naive, and it seemed the older man had taken him under his wing. It was midafternoon, and they were drinking beer. The older man was talking confidently about something, and he stopped when he turned toward me. He called out as I was walking past and asked what I was doing. I told him that I had just spent a week on foot out in the oil field.

"So," he said, "did you have any big epiphanies on your trek?"

His question caught me off guard. At home in New Hampshire I spent my nights in confident comfort. The prairie was reeling, and so was I. Louise and her husband, and the people in the campground and clamoring all around, needed the work. That is the story of this nation and the world: growth is progress. And wasn't all the effort in the oil field being done for people like my family and me: upending energy to keep me comfortable back home in New Hampshire?

The climate crisis demanded an end to fossil fuels. But in North Dakota, the old industry of oil was digging deeper, racing to keep crude at the center of the system. And the Bakken oil field, for all its miles of open sprawl and thousands of wells reaching into the earth, was just one small piece of the global oil infrastructure: from the north slope of Alaska to the Permian Basin of Texas and salt-water stretches of the Gulf of Mexico, from Venezuela to Nigeria and the Middle East.

I did feel more alert from exposure to people and landscape during my walk, and my steps from Tobacco Garden had traversed a longer arc of time than the man at the picnic table might feel day after day in the campground. But I was overwhelmed and exhausted from my adventure. I balked at his question about epiphanies.

"Not yet," I said, turning away.

I would wander into town the next day, and I was unsure whether I would find calm or chaos. That night I waited until dark before crawling into my tent.

※

At nine in the morning the first five pews in the Lutheran church one block off Main Street were still empty. Pastor Mark Honstein, robed and wide-smiled, strolled back to the sixth pew to stand close to the Sunday worshippers. He had a few announcements. The food pantry would be open Tuesday afternoon from five to seven o'clock. The parish's country churches were switching to summer hours. The Lutheran quilters would meet Monday afternoon, as they do. And coffee hour after the church service would be held outside on the lawn, given that it was such a nice day. Pastor Garrett Gudmunsen could not be in church to give the sermon, as scheduled, because he was out on a ranch branding calves. "That's where he is today, if any of you want to join him," Honstein told the worshippers.

The congregation tended toward the older population, men in collared shirts, women in skirts. They had arrived in slow-moving sedans and pickup trucks that navigated familiar turns of the city grid and then parked only a block from a pharmacy, which still had a soda fountain in back. The brick steeple of the downtown church was more formidable than those in the countryside. The sanctuary was set beneath wide wooden beams. Honstein worked to overcome that formality. "Out here, farmers, we help each other," he told the congregation. "Ranchers, we help each other. But at the same time you have to keep your own business running." Honstein mentioned Matthew, the tax collector; Simon, the zealot; and Judas, who handled the money. "What brought them all together? They were together in Christ," Honstein said.

The congregation sat still, like the brick steeple overhead, and Honstein knew that the gulf was great between those in his church, long rooted in this place, and the rising sea of newcomers. There were plans in Watford City for a new high school, a new hospital, and new housing developments to cater to the oil-field crowd. A sprawling shopping center that opened the year before had shifted Watford City's commercial core from Main Street to a

mile south, alongside the highway thick with oil trucks. Drug traf-
ficking and domestic violence were overwhelming police. Prices
for nearly everything were rising month after month.

"This world," Honstein said, "is a hard world to live in." He
turned again to the Lord that the Lutherans had brought with
them to the prairie a century ago. He urged the congregation to
have empathy for the new civilization emerging. And he talked
of the tornado that touched down earlier that week. He told the
worshippers that he wanted them to hold the tornado's victims in
their hearts that morning. "We are together in Christ, and together
with Christ," Honstein said. "That is the purpose of suffering."

One hour later, down past the new shopping center, pickup
trucks weary from oil-field work rested in a wide dirt parking lot.
The building of the Assembly of God Church was itself a work in
progress, with a large wing under construction to welcome the
expanding congregation. Inside the lobby kids darted between
legs in the crowd, and one man explained to another that he had
driven more than an hour to get there. "There's a lot of need for
Jesus here," he said. In the sanctuary a band onstage was rocking
out in the name of the Lord amid colorful cardboard in the shapes
of a garden. Lyrics were projected onto an overhead screen, as the
lead guitarist sang, "I have decided to follow Jesus / No turning
back / No turning back."

Seated among the newcomers were some longtime locals, including the mayor of Watford City, Brent Sanford, who rose from
his seat and strolled to the stage. He told the crowd what had gone
through his mind when he had heard the tornado tore through
an RV park. "Which one? Where at? How many people were injured?" Sanford said. And then: "I truly believe a miracle occurred, to have a tornado touch down among so many units, so
many thousands of people, and no one was killed." A fifteen-year-
old girl from Louisiana remained hospitalized in Minot in critical
condition, but seven other people who were injured had all been

released. "Half of them are heading home," Sanford said. "The other half are looking to get back on their feet."

Sanford sat down, and Pastor Sheldon McGorman set into prayer: "More than anything, I want to thank you for your grace and your mercy that spared people's lives this week, God, that your hand was on our community. God, we give you all the praise." The band was still onstage, and the guitarist began to slowly strum as McGorman continued. "God, you literally pulled people from the funnel cloud, and they were saved and spared, God," he said. "We thank you, Lord Jesus, for doing that in our community." McGorman's voice rose in urgency. "We thank you that this region is a special region, that this region is dedicated to God, that this region was created so that people could come to know Jesus," he said. From the crowd came murmurs of agreement, and even up in the balcony McGorman's spoken words resonated: "It is a special time frame that you have designed, God, for people's lives to be touched and changed by you."

※

On the low hill south of Watford City, where the tornado had touched down six days before, it was clear the natural order had been broken, and the place, accessed by rutted road from the paved two-lane highway, had the feel of terror stilled. Twists of steel were piled atop dirt. Sheets of metal were crumpled like paper, and broken wooden beams rose at odd angles, bent nails clawing at the air. I spun in the bright light of the prairie, feeling not cradled in a bosom of love and protection, but exposed in a hardscrabble landscape of individual survival.

A few feet away from the uprooted RVs, other camper trailers stood unscathed, and pickup trucks were parked at their doors. A car pulled up to a low building that held one-room apartments, the doors of many of which were open. A woman in front of one door stood in shadow with a baby on her hip. A toddler stumbled along a concrete walkway that was near an open pit, from which a man emerged. The man's head was close-shaven, his gaze a disarming mix of hardness and care. His name was Derrick Wilson, and he was working on plumbing for the building, where he lived in an apartment with his eighteen-month-old son, Aidan, and the boy's mother. "It ain't no life living in these campers," he said, with a stiff nod to the neighboring lot.

Derrick, who was thirty-seven, had worked in oil and gas fields for eighteen years, from New York to California, Texas to Canada. "Oklahoma's where I'm from," he said. Derrick had seen the tornado as it arrived from the west a week before, skidding above open pasture on the other side of U.S. Highway 85 and then jumping the road and touching down in the thick of the RVs. He had shouted and shepherded unknown neighbors to safety in the low apartment building in those final minutes. Standing again beneath blue sky, he told me he was not hopeful that many of those perched on the prairie had found a permanent place. Everything there relied on global economics. If the price of a barrel of oil trading on distant markets should drop, and it always does, eventually, then the companies rushing to drill wells and dig pipelines would pack up just as quickly as they had arrived.

Derrick knew that need brought people to work in the Bakken, and if the drilling stopped, many would move in search of another opportunity. He had little confidence that in the meantime Watford City leaders could provide enough housing, classrooms, police, or social services for the people crowding around town. "By the time they get their stuff worked out," Derrick said, "the boom's gonna be over."

The tornado had stopped short of a neighboring collection of tall sheds, in which hundreds of oil-field workers sought more protection. A spot in the Indoor RV Park, as it is known, cost $1,200 a month, nearly twice as much as a spot exposed to the elements. The high price brought relative comfort inside the sheds:

residents parked their trailers in a space that stays cool in summer and warm in winter. Batteries were not so quick to die, hoses not so likely to freeze. Each shed opened with a high garage door, and inside the third slot of the farthest shed that Sunday afternoon Tomi Ballard stood sturdily over an ironing board, her back to the camper trailer that she lived in with her husband, Aaron.

Tomi's eyes beamed brighter than her fifty-five years, and she glanced from the board to a makeshift rack holding workers' clothes. She told me that after a second factory had closed in Arkansas, costing Aaron his job, the couple packed up their belongings from a two-bedroom home on two acres of land and rolled out to North Dakota. "We landed here and started looking," Tomi told me. "We had nothing. We had a little bit of money in our pocket."

Aaron soon found a job helping to build a gas plant, and Tomi cleaned neighbors' RVs and took care of their clothes. "I invested me in a good iron and ironing board," she said. The welders, especially, wanted their clothes crisp, and Tomi found a spray starch that almost did the job. That afternoon Tomi was hoping to iron thirteen sets of work clothes and another seven pairs of pants. "No, I cannot make them as stiff as a dry cleaner," she said. "But I can get them close."

Tomi and Aaron had built wooden steps leading up to the door of their camper and set on the railing a hand-painted sign in the shape of a running razorback pig. They had rolled outdoor carpet on the gravel of the shed floor and arranged two card tables, one of which held a framed black-and-white photo of Tomi's parents. "This year, there's a lot more women than when I first got here," Tomi said. "We've got five couples in here that we truly believe to be friends."

On Memorial Day evening she had been gathered with many of them when they saw the tornado move in from the west. She described to me how the gray funnel cloud flickered with red and blue and green as the RVs were lifted into the sky. It was enough to rattle Tomi and Aaron, but not so much that they would leave. The shed had become home: the couple's son-in-law was planning to move from Arkansas with the grandkids. "My babies," she called them. As she talked with me, Tomi's cell phone rang. The man on the phone let Tomi know her family could park another RV in the neighboring space. "Love you, Mr. Joe," Tomi said into the phone. "Thank you."

Sitting at Tomi's table, I felt oddly at home in the uncertain place. Tomi and others had welcomed me among them easily, sharing the intimacy of the uprooted. But I knew that it was a false comfort, as they and everyone in the oil field existed atop a fragile infrastructure that could collapse at any time. The lives of those who came to the prairie seemed to me a symbol for the bigger threat to the planet: so many people digging too fast to claim more oil and feed a system already burning too much. That rush took a toll on those who work in the oil field, to be sure, but I knew it threatened everyone. My daily life in New Hampshire was more stable, with a roof over my head and the support of a more settled community. But that insulation from the source had only left me ignorant, unaware of the natural risks that such a sheltered life creates.

After so much movement across the prairie, I was tired and unsure of my next steps. In a few days I would begin my journey back home to my family in New Hampshire. Then, two months later, I would walk again, heading into the gas country of New York and Pennsylvania, where I hoped I would learn more about the larger stakes of the fossil-fuel system.

Tomi had begun our visit by offering me sweet tea. I drank that, and Tomi invited me to return the next evening to meet the neighbors and enjoy home-cooked barbecue ribs and all the fixings. First, though, she told me I really ought to go meet Owen and Felicia Grooms. The young couple had lost nearly everything in the tornado and had taken shelter at a hotel in Watford City.

※

Just west of the main intersection on the south side of Watford City, where oil-field traffic backed up, the low-slung Roosevelt Hotel seemed almost lost behind an overgrown bust of Theodore Roosevelt that dominated the parking lot. A few minutes after I arrived in the cramped lobby, Owen and Felicia Grooms came down from their room to meet me. Their two-year-old son was napping upstairs. Owen took a seat, and Felicia stood. They looked calmly bewildered.

Owen explained that two years earlier they had sold four of their six horses in Idaho to raise money to take a chance in the oil field. They had planned to earn enough working in the Bakken for a down payment on a house of their own back in Idaho. "We come from mountains and trees," Felicia said.

During their two years in North Dakota, the Grooms had paid off debts and begun saving for a down payment. As Owen talked, Felicia braced behind him. "We were almost done," Owen said. But when the tornado descended on the RV park, the young family lost their trailer and everything in it.

Time was turning quickly for Owen and Felicia, and they knew they had to react. The Roosevelt was offering a free room for some nights, and the couple hoped a local fund for tornado victims might provide enough to find another trailer and get back to work. I asked if the couple's son was aware of the tornado and how much it threatened their place in the prairie. Felicia, exhaustion in her eyes, told me that the boy had lived his whole life in that trailer. In the hotel, she said, he could sense their vulnerability. "He definitely knows," Felicia said, "that we are not home."

# Chapter 2
# States of Combustion

Marie McRae is strong of frame but soft of feature, her voice firm, but often with a hollowness that suggests she would rather not be talking at all. When I drove into her farmyard in New York State on an August morning, two months after walking across the North Dakota prairie, a pile of firewood was waiting to be stacked and dried for use in two winters. "My tap root goes really deep," Marie told me.

Six Mile Creek ran clear alongside the pasture behind Marie's barn. Horses in a small corral near the farmhouse swatted their tails at flies. Bees buzzed against window panes in humid sunlight. Indoors, a fresh-baked pan of pumpkin bread cooled on the kitchen counter. Marie took a seat at her living room table, shadowed against late summer sun. She drew a hesitant breath and talked about a day six years earlier when another stranger had appeared at the screen door of her farmhouse. Marie repeated the stranger's words: "This is your last chance to have a say about what happens on your land, because we're coming in."

In that fall of 2008 many of Marie's neighbors up and down the valley of fields and forest in southern New York had already leased land to companies hoping to drill for natural gas. A gas-fracking boom had taken off forty miles south, across the border in Pennsylvania. There was, at the time, a statewide moratorium against hydraulic fracturing in New York, but powerful corporations were lobbying to overturn that. And the land man who came knocking on Marie's door told her that drilling would soon surround her farm. "And so I signed," she told me. Her voice weakened in the memory. "After that, I mentally crawled up in a ball."

Marie did not stay crawled up for long. Within months of signing the lease, she joined others in and around her town of Dryden who were organizing in living rooms and law offices. They gathered support of 1,600 Dryden residents, and the town board agreed to ban drilling, a proposition built on New York's policy of home rule.

The fight against more drilling by Marie and others who live in the modern world and consume its resources could be seen as a NIMBY move: Not in My Backyard! But in staking a claim against fracking in the forested ridges of southern New York, they were challenging the powerful inertia of society's collective consumption. Like Don Nelson, the North Dakota rancher whose land already had several oil wells on it, Marie sensed at a gut level that pumping chemicals deep into the earth to extract gas would upset the natural balance. "If you take a bore, and you go down a mile," Marie told me, "this is all I know: You're going to damage the water."

At the time of my visit, New York's moratorium was still open for debate, and it was unclear whether Gov. Andrew Cuomo would make the fracking ban permanent. Marie followed developments across the border in Pennsylvania, as drilling and the pipeline infrastructure it requires redefined rural terrain. She talked about the need to preserve nature rather than claim more of it to keep pace with industrialized life. "It is the war," she said, "that is before us."

It felt good to be sitting with Marie in a place where people had decided to say no to the harvest of more fossil fuel. After all the action I had seen in North Dakota, I was eager to walk next through terrain that would offer deeper consideration of the collective decisions made to support the industrial system. And I was hopeful that my seventy-mile route, heading south from Marie's home along quiet New York roads and then into another drilling boom, in Pennsylvania, would offer a second stage to my reckoning of fossil-fuel landscapes. The fuel in this place was different: natural gas instead of oil. But the same technology of horizontal drilling and hydraulic fracturing that created the Bakken boom also made possible the harvest of gas long-compressed inside the Marcellus shale, beneath New York, Pennsylvania, and much of Appalachia.

New York's moratorium against fracking had left the gas industry and some locals chafing for action. Across the border, in the Endless Mountains of Pennsylvania, a decade of drilling had damaged water wells and brought widespread industrialization to the countryside—from buried pipelines to hissing compressor stations to polluted air. I was confident that in the days ahead I would have a chance to measure this older boom's deeper cost, for the place and the planet.

Marie turned our talk from questions of energy and the environment to the personal toll of modern life, in which people live so separated from the source of their survival. Even in Dryden, she said, young people grow up at a suburban pace, disconnected from the land beneath their feet. She described how kids drift for meaning as they come of age. She worried at the lack of daily tasks, such as chopping wood and feeding chickens, that directly relate to survival and comfort. "There's no feeling of purpose," she said. "There's no feeling of 'right work.' And if there's no right work for you, then where are you? You are not grounded in the universe."

I realized that my walking—shouldering a pack and sweating through landscapes normally traveled by others with motorized ease—was becoming a kind of "right work" of my own. Slowing my pace, exposing myself to occasional risk and frequent discomfort, was helping me physically connect to unfamiliar terrain. Moving step-by-step, feeling the contours of the land and encountering people working and living on it, was helping me overcome the deep dislocation I felt in my insulated life at home in New Hampshire. I was coming to understand both the scale of my consumption and the danger of a society digging deeper for more fossil fuels as the climate crisis accelerated.

The poet Alfred Lord Tennyson once wrote, in the words of his character Ulysses: "I am a part of all that I have met." Tennyson meant that Ulysses carried with him all he had experienced during travels in far-flung lands. People living in the industrial era are, through their consumption of fuels from faraway landscapes, parts of places they have never been. I had seen in North Dakota that my years spent burning oil—while driving a car or riding in an airplane—had made me a part of the prairie and the oil harvest there. Ryan Bak, the oil-well maintenance worker who had asked me who would give up a big-screen television or other conveniences and comforts oil brings, had made that point more than clear. And as I arrived in New York and Pennsylvania, states that sat above the deep stores of Marcellus natural gas, I knew my home in New Hampshire counted on such fuel. The gas-fracking boom in Pennsylvania was the very thing that made the new gas furnace on my basement wall affordable.

My journey on foot through North Dakota had been all open

prairie and big sky, and I had found there an energy in my movement, the vastness giving purpose to each step I took toward the
horizon. The route I planned to follow south from Marie's farmhouse would traverse markedly different terrain: thick-leafed
oaks and green-needled pines cloaking hills and hamlets all along
the way. One local in southern New York I had spoken with weeks
earlier by phone had described the landscape as two-dimensional,
framed at every turn by the rise and fall of elevation or a shield of
forest. "You can only see," he told me, "what isn't blocked."

As my conversation with Marie turned toward a second hour,
she grew restless. She was eager to get back to a renovation project she was working on at a house down the road. She cut me two
big slices from the pan of pumpkin bread and walked me to the
door. I stepped onto the short stoop. To the west, black clouds
crowded green treetops. I hoisted my pack. Low thunder rumbled,
as though two neighbors were debating what would happen next.

⁂

With my first few steps onto Hammond Hill Road, which runs
south from Marie's property, dark forest closed tightly against
smooth asphalt, and the road began a steep climb. The weight of
my pack was pulling me back, and I had to lean into each step just
to keep heading forward. The road angled around a bend, and neither the grade of the climb nor the tightening of the trees eased.

I huffed for ten minutes and more, straight uphill. There were
no switchbacks, as might be found on a footpath, because this
road was built for cars and trucks that had more horsepower. Suddenly, though, Hammond Hill leveled flat, and pavement gave
way to a single-track dirt lane, which continued straight into
denser woods, branches closing low overhead. Thunder had been
steady as I climbed the hill, and by then it was pouring rain and
nearly dark as dusk beneath the trees. I stepped from the road, already turned to mud, to take cover under a towering pine. Thick

drops fell as mist. I pulled out a garbage bag to cover my pack and lowered myself against the rough trunk.

Hammond Hill is just one rise on the Allegheny Plateau, a swath of sedimentary rock pushed upward during the collision of the North American and African continents roughly 300,000,000 years ago. Retreating ice sheets 299,990,000 years later carved much of the topographical terrain that exists today, and that has been weathered steadily since, polished with constant rainfall, such as that dripping on me beneath the tree. When the first European explorers ventured into the region, long inhabited by native peoples, they were so spooked by the density of woods that they called the place "dark forest," as though it were a feature on a fairy tale map. During the century after the creation of the United States, most forests in southern New York were cleared for farmland. Trees were used for timber or potash, a valuable fertilizer, but much of the soil didn't grow good crops. So over the most recent century, particularly between the 1930s and 1950s, government crews pushed seeds back into the earth. South of where I sat, more than two million seeds of red pine, white pine, Norway spruce, and Scotch pine had been planted during a few years.

The modern forest, woven with paths for horseback riding and hiking, hunting and holiday picnics, is a recreation of an ideal already undone. I knew, as I sat alone and waited out the rain, that the real value of this place, in any sense that the modern world had come to cherish, lay in the gas-rich Marcellus shale, three thousand feet beneath the surface where I sat. As I walked in the coming days, that distance would grow, as the layer of shale slants deeper underground toward the south.

The rain began to let up, and I slopped along the dirt track atop Hammond Hill. Deep puddles filled ruts as wide as the lane itself. My boots sank into soft earth as I planted the stick and pivoted at the edges of deep pools, and there I often bothered frogs. They sat on logs and branches, sentinels at their posts. When I splashed, they dove deep into the water, a reminder of how much lives in it.

Late that afternoon, on a forested stretch without homes or commerce, I came to a scatter of buildings surrounded by a high fence with barbed wire at the top. A sign announced that the place was the MacCormick Secure Center, a juvenile prison. I had run out of drinking water, so I sat on the ground and waited for someone to emerge. The MacCormick Secure Center was surrounded by Shindagin Hollow State Forest, and I'd been planning to head into the woods and pitch my tent. It began raining harder, a gloomy dusk settling in, and I debated moving toward camp. Before I did, a pickup truck pulled into the lot, and a man climbed out. He said, "Follow me around back to the pump house." He drove his truck, and I walked behind. As we traced the fence line, I saw the prison was quite small, the size of a country school. I stared through the chain link at an empty yard and a warren of one-story modular buildings.

I filled two bottles from the pump-house sink. I asked the man who was being held inside the modular buildings. He said that MacCormick housed young people who had committed violent crimes. I stared across the yard at the barred windows in each building, and I told the man I could not imagine that a troubled teenager had much hope for recovery in there. He looked at me a long time, as though preparing a soliloquy, then said, "It's just a sad thing."

A quarter mile down the road, I turned into the ditch, then up a steep slope into the woods of Shindagin Hollow. The forest stretched for miles to the west, but I stayed close enough to the road that I could see vehicles passing every few minutes. I picked a spot between two towering pines, their canopy overhead easily a hundred feet high. The soft mud at my feet was covered in decomposing needles and bark. Nearby a falling trunk had wedged high in the branches of a neighboring tree, so that the trunk rested at a forty-five-degree angle, paused in its collapse. The rain had slowed to an occasional splatter. I pitched my tent quickly, and I rolled all the clothes I'd been wearing into a trash bag and put on dry pants and a shirt. I spread out my sleeping bag and climbed inside the tent.

I could see through the screen a toad hopping only inches away. Was he trying to join me, or had I set down my tent on his home? I stared around him at a small swath of ground rich with elements: leaves, twigs, mushrooms, bark, moss, shoots of new trees, squirrel tracks, spiders, mosquitoes. It was all wet and interwoven. I wondered briefly about bears. My muscles ached after that first long day. Darkness came, and I soon slept.

In the dead of night I woke to thunder. Rain was pouring down harder than it had during the day. Lightning flashed and the nylon all around me was visible for a second, or two, at a time. I thought of the children locked in the buildings up the road. Society is so good at walling itself off from things it does not want to see. I thought of the Marcellus shale buried thousands of feet beneath me and of the drilling I would encounter a few days later when I crossed the border into Pennsylvania. I thought of my wife, son, and daughter sleeping in the surety of our home in New Hampshire. Their separation from the source of their survival so often seemed a comfort, but I knew as I lay exposed in the storm-struck forest that it came at a deeper cost.

One year before I walked above the Marcellus, a measuring device atop Mauna Loa, the Hawaiian volcano, documented what scientists saw as a crucial number: 400 parts per million of carbon dioxide in the atmosphere. Until that March day in 2013, that threshold had not been crossed for millions of years, before the existence of humans and at a time when the planet was several degrees hotter and sea levels were thirty feet or more higher. The National Oceanic and Atmospheric Administration scientists monitoring the instruments noted that the carbon dioxide had jumped in just six decades from roughly 280 parts per million, which had been constant prior to the Industrial Revolution, to the 400 mark.

That sixty-year span matched a wild increase of carbon emitted by humans, largely through the burning of fossil fuels: the amount of carbon dioxide released each year increased from five billion tons in 1950 to thirty-five billion tons in 2014. At news of the 400 parts per million measurement, Annmarie Eldering, a lead scientist with NASA's Orbiting Carbon Observatory-2 satellite project, had joined other experts warning that such high carbon levels would only accelerate global warming and climate change. "Passing this mark," she said, "should motivate us to advocate for focused efforts to reduce emissions around the globe."

Surrounded by the soaked sounds of Shindagin Hollow, I lay awake in my tent in that darkest hour of the storm. I knew that society needed, as Marie McRae had done, to demand another course forward. The thunder broke directly overhead. Each clap echoed in my chest, and I stretched long against the shuddering soil.

❋

Down into the valley along Honeypot Road, the ground was warm and the air cool after a night of rain. Dense clouds seemed to hold the terrain in a timeless state. A red barn lurked at a bend in the road. Fields of grass lay flat beneath the weight of water. The road had two paved lanes that weaved between those fields, dropping from the Shindagin Hollow forest more than seven hundred feet in five miles, a steep enough descent with wheels and a pure pounding foot by foot. It ended at a stop-sign intersection with two more lanes, these of the wider, busier Route 96, which ran between Spencer to the west and Owego to the south. Nestled along both sides of Route 96, just past the stop sign, sat the village of Candor, born in the westward push after the American Revolution and home to 850 people. At eight o'clock on a

Saturday morning, the streets were quiet, front doors shut and curtains drawn against the soon-to-rise August heat. The only action appeared at the corner of Main and Owego Streets, where the XTRAmart offered gas pumps outside and aisles of snacks inside: cheap fuel to keep moving. I slumped against the front of the store and waited until a Subaru Forrester pulled into the lot, its wheels caked in mud, its doors splattered with the same, and out stepped Bob Aman.

Bob had wispy gray hair and broad shoulders, with a forward slope to his body and face that gave him a look of both strength and softness, a range of character that might come from having lived six decades in the place he was born. His grown son, Aaron, sat in the front seat, and both men clutched big cups of coffee that added to their energy. The Amans run one of the few dairy farms left in a valley that used to be home to more than thirty, and I'd come to meet them because other people had told me that Bob, like many in the town, was in favor of fracking for natural gas. I hopped in the Subaru's backseat, and Bob drove a few blocks south to drop Aaron at home. Bob and I would talk a lot that morning about the history of gas in the valley and the potential for more development, but Aaron chimed in first, saying he hoped the fracking would never cross the border from Pennsylvania into New York. As he climbed out of the car, Aaron turned to me with a quiet voice. "The issue is divisive even within families," he said. "If it comes, it comes. If it doesn't, it doesn't. I'm going to keep farming."

I took Aaron's place in the front passenger seat, and Bob aimed the Subaru north again, driving for a few blocks along a central street lined with two-story houses and small yards that had been built when Candor was a center of its own. Native Americans had been living in the area for thousands of years when, on another August day, in 1779, revolutionary troops followed the rivers into the area around modern-day Owego, ten miles south of Candor.

Ah-wa-ga, the settlement was called in the Cayuga language, and, in a strike against the British allies, the soldiers burned the Iroquois lodgings. The glimpse of the thick forest the soldiers got during those attacks lingered, and after the Revolutionary War ended, soldiers claimed settlements around present-day Candor as the young country grew west. Families with names such as Brown, Bacon, Hart, and Draper made the two-hundred-mile journey west from Connecticut and other seaboard states. Timber soon was falling fast, and lumber was shipped south on the Susquehanna River to the Chesapeake. By 1834 Catatonk Creek, which runs through the center of Candor, was dammed, and Jesse and Ogden Smith built their gristmill on Upper Mill Pond, at the north edge of the village.

For more than a century, newcomers found jobs at a blanket factory, a woolen mill, a leather tannery, and other enterprises. The Lackawanna railroad, which had a depot in town, connected products to distant markets. By the count of Carol Henry, a local historian, the village and surrounding area were producing, in the late 1800s, thirty thousand yards of cloth per year, three hundred bushels of wheat per day, and nearly three thousand sides of leather per month. Then coal and oil came to replace water power, and, with the rise of the automobile and the expansion of industry elsewhere, Candor calmed during the twentieth century. The train tracks were abandoned in the 1950s, and small family farms collapsed in the decades after. By the day I met Bob, the town had long since turned into more of a bedroom community for people employed in Ithaca and elsewhere. Saturday morning was a chance for rest.

Bob turned east a few blocks north of the XTRAmart, and then north again, past some low buildings surrounded by a field. We had been talking only a few minutes, more about life in town than the potential of fracking for natural gas, but Bob quickly expanded our conversation. The global population, which counted

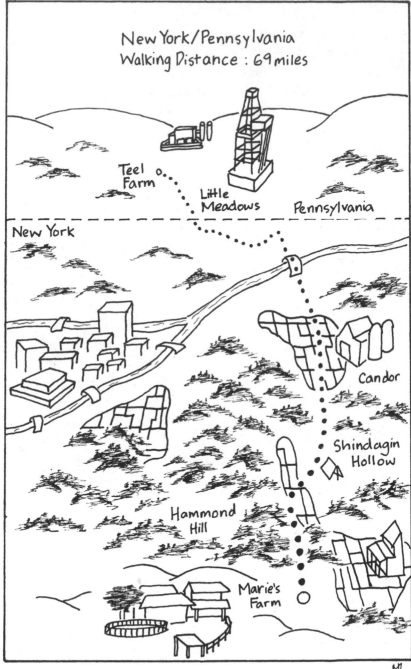

1.6 billion people in 1900, had leaped to 6.1 billion by 2000 and crossed 7.0 billion by 2012. "Eventually," Bob said, "there will be so many people that we'll wipe ourselves out, some way, shape, or manner."

We pulled to a stop next to a low pool, roughly the size of a football field, that was full of cow manure. Thick black plastic lining kept the manure in place, as the pool was filled from a network of pipes that Bob had dug beneath the village, all of it carting a daily deluge of manure from his 550 cows. The cows produce roughly four thousand gallons of milk each day, but more than three times that much manure. Manure is full of bacteria and methane and is, alongside fossil fuels, a major contributor to global warming. But it also has potential for power, and for more than a decade Bob had been pumping poop through what's called a *digester*, where it is heated to nearly a hundred degrees, releasing the methane gas, which is harnessed by a diesel engine to produce electricity. Bob saved an estimated $3,000 a month powering his farm operations, but also pumped some of the electricity back into the grid. The system had a simpler benefit, more detectable to anyone in town: it cut down the odor of manure, which had previously only been spread on fields.

For all his tinkering with manure and its methane, Bob was also eager to see the fracking industry arrive to unlock natural gas buried in the Marcellus deep beneath his farm. The natural-gas industry got its start two centuries earlier, just two hundred miles away, in western New York, where a man named William Hart dug a well at the site of a bubbling gas spring in a creek, then piped the gas around the town of Fredonia. By 1825 stores and a mill site became among the first in the United States to be lit by lamps fueled by gas. For more than a hundred years, a lack of pipeline infrastructure limited gas's use mostly to lighting for local customers. But during the past century, the potent methane, which holds the carbon in the gas, has increasingly been burned to power

electricity plants. A vast network of thousands of miles of underground pipelines traverses the United States like so many veins.

Most early gas, including that in Fredonia, came from underground pockets wedged between layers of rock. But the combination of horizontal drilling and hydraulic fracturing that opened
up oil in North Dakota and elsewhere has done the same for natural gas, making the so-called tight gas formations in the Marcellus shale a target. The natural gas–fracking boom in Arkansas and
California, Texas and Colorado, had increased natural-gas production in the United States by more than 50 percent in the decade
before I walked into New York. Across the border in Pennsylvania, during eight fast years seventy companies had drilled a total
of more than five thousand wells into the Marcellus shale, and
Bob thought New York should see some of that action.

He drove a few miles west of Candor, toward his dairy barns.
The valley was modern rural order: close-cropped fields and forest on the ridges. Bob turned onto a long drive and crossed a mud-
slop lot. He stopped near a series of fuel tanks that held diesel,
gasoline, and kerosene. The tanks were filled regularly by visiting
trucks to keep farm operations humming. Even with the manure
system, the farm still required fossil-fuel harvest and delivery.
And Bob worried about what would happen if that system were
to fail, even briefly. "If we don't get fuel, our trucks don't get fuel,"
Bob said. "And if the overland trucks don't get fuel, you've seen
the studies about how quickly store shelves go empty," he said.
"Sometimes within a day."

Bob and I climbed back into the car, and he turned onto a road
heading west toward a plot of his land where Talisman Energy had
drilled an exploratory gas well years earlier. It was a conventional
well, with a pipe descending eight thousand feet into the earth.
While it passed through the Marcellus and its tightly packed
stores of gas, the Talisman well did not hit any easily accessible
pockets. So by the time Bob and I arrived at the site, a concrete

cap covered the scene of the effort, and a barren patch of dirt was surrounded by stalks of midsummer corn.

Hopeful that fracking—with its horizontal drilling and powerful pressure of water and chemicals—could tap into the Marcellus, Bob had joined a group known as the Tioga County Landowners Association. The landowners wanted to be in position to negotiate good lease rates for their property if the fracking boom were to move north from Pennsylvania. "We all got together, because we kind of knew that we weren't capable of keeping up with the gas companies," Bob told me. "They were writing the leases to their benefit, no question about it." But New York's fracking moratorium, which had been in place for five years already, prevented any new natural-gas drilling before it began. And the summer after I walked through Candor, the state would make the temporary ban permanent, citing risks for land, water, natural resources, and public health.

Bob had to pick up his grandson, Aiden, from football practice, and we motored back toward town. He parked behind the school, and Aiden climbed into the backseat, his pads and helmet resting on the seat beside him. With 2,200 acres of dairy farm, Bob stood to profit handsomely if fracking were ever allowed. That would benefit Aiden and others of his generation in the family. But Bob didn't want Aiden to have so much money that he could survive without exertion and effort. I rode with Aiden and Bob into the center of Candor, where he dropped me at the curb. Before I climbed out Bob explained that he did worry that a gas boom, and personal profit, could separate Aiden from his purpose on the land in the valley. "I want him to still have to work," Bob said.

It was already after noon, and Candor had come to life with the rituals of keeping pace during a break from the workweek: Residents running errands to big-box stores in Ithaca or taking time to tidy the yard. My tent and gear dry from a morning that mixed clouds and sun, I loaded my pack and began the ten-mile walk south to Owego. The land less open than out West, the routes more rutted along the valley floors, I had little choice but to follow the two lanes of Route 96 as it snaked southward, tracing the banks of Catatonk Creek.

The roadside soon returned to field, and I kept pace for one mile after the next. The fields were broken only by occasional homes and, every quarter mile or so, some kind of enterprise: Warner's Stoker Stoves & Coal, Brush & Palette Auto, Double Aught Lumber, Route 96 Power & Paddle, Dollar General, Hollenbeck's Feed & Coal, Catatonk Truck Sales. Cars and trucks whistled past, heading north and south in flocks of threes and fours. Their engines hummed at rest after a quick acceleration or growled deep as the driver pushed foot to pedal, some vehicles holding a tight course, sturdy on their wheels; others weaving a bit from too much speed or driver's distraction. Walking just a few feet outside of the lane, I felt every tweak and shudder as the passengers swept past. Moving step-by-step, I knew that, despite farms and forest all around, this was no kind of natural at all.

I turned from the road two miles south to explore a haunted house that came to life every October. Then, at a Baptist church seven miles south of Candor, I unlaced my boot to examine a burgeoning blister. As I took more steps along Highway 96, I passed above the Millennium Pipeline, buried underground. That pipe runs from the western end of the state, where traditional drilling has long produced natural gas. It carries the gas eastward to the town of Ramapo, where it connects to the Algonquin Pipeline, which angles north into New England, powering electrical plants and homes for millions of people.

As I hobbled the final few miles into Owego, I stopped on the shoulder of the road and stared down at the carcass of a dead bird, hit by a passing vehicle. Such roadkill, I'd come to realize while walking, was also a barometer of the imbalance: other species, shuttling about in their daily lives, struck down from the velocity of movement. The bird's brown feathers froze in folded wing, its brittle legs angled as though poised for an uncertain step. I stood still, as cars and trucks passed just a few feet away. The scene at my feet reminded me of one I'd encountered while descending the Honeypot Road from Shindagin Hollow toward Candor. There I had come across a painted turtle squashed on the asphalt. Its shell, which still held its natural shape, had been shattered into several pieces.

※

While walking, I had met a retired man who invited me to join his friends for their morning coffee, and early the next day, in the village of Owego, I stopped at a bend, where a farmhouse hugged close the road. Around back in the kitchen, nine men sat in a scatter of chairs, coffee mugs in hands, quips flying from lips, idling into another day. The group called themselves the Squirrel Club. "Because we're nuts," one man said with a laugh. A collection of them—which that Sunday included Skip and Spencer, Andy and Bryce, Benny and Dick—gathered each morning, as though a fast-forward scene from elementary school, the kids still together, though nearing their ends. "We were all the ones that were two goddamn ignorant to leave," said another man, a gray T-shirt covering a belly that matched the boom of his voice.

But there had been much to keep these men living in the hills and hollows around Owego, on the banks of the wide-flowing Susquehanna River. Two of the men were born and grew up a mile apart. Another was raised on a nearby farm that had been tilled by his family since the late 1800s. Said a fourth man: "I was born in Owego, lived in Owego, and I've only got to go a quarter mile to finish." He playfully jerked his thumb in the direction of the village cemetery.

The kitchen in which the men gathered had faux wood paneling and a linoleum floor, dating at least to the 1970s. A dish of cinnamon apples sat on the folding card table, and a red plastic tub of instant coffee held fewer crystals than it had an hour before. Conversation ebbed and flowed, and debate kicked up about whether the county sheriff's deputies should be more ambitious and whether village police were even needed anymore. Owego's population had dropped from 5,400 to 3,900 in the course of a few decades, primarily because the IBM facility in Binghamton cut thousands of jobs. Local real estate had fallen 30 percent in the three years since an epic flood washed over Owego's streets. "The village is broke," one man said. "No money."

Eventually, one member of the Squirrel Club turned to me and asked why I was traveling on foot. I told the men about my trip to North Dakota's Bakken bonanza and that I would cross the border later that afternoon into Pennsylvania and its gas-fracking boom. In New York, I said, I had wanted to see a place not yet hit by more fossil-fuel development. "So you're writing about that: the lack of," one man said. Yes, I told him.

Several men in the room criticized the fracking moratorium and said they would welcome natural-gas fracking in and around Owego. It would, they said, bring lease revenues for landowners and jobs for young people who too often take flight. "I think New York shot themselves in the foot," one man said.

Another man was confident that society's appetite would be too great to prevent drilling forever, and the Marcellus beneath Owego would be drilled too. "It's going to come; it's going to come," he said. "But it's a long ways off." Someone else chimed in: "After we're gone." There was a clear sense among them that, no matter the issue, time was ticking fast, and an animal instinct to keep moving forward required more action. Even their homes had not held value, offering little financial certainty in retirement. "I paid taxes for fifty years, hoping some day it would amount to something," one man said.

I needed to get walking if I were going to make it twelve miles that afternoon to Little Meadows, a village just across the border in Pennsylvania. The men told me I was welcome to drop in again, as they gather for morning coffee 364 days a year, a rhythm almost without end. "You can come Christmas, Easter, New Year's," one man said. "But the first day of deer season, no one's here."

As I hobbled south that afternoon, having crossed the Susquehanna River soon after leaving the Squirrel Club farmhouse, I followed Chestnut Ridge Road until it twisted on a steep descent into a valley, where it met Pennsylvania Avenue. That rose only slightly, one mile after the next, as it traced Apalachin Creek, which flowed from the south, beginning its journey five miles over the border in Pennsylvania, near Minkler Lake. The lake was more like a pond, judging by my map, but just downstream from that the Apalachin gained current after the confluence with Bow Bridge Creek, which springs from the ground deep in the trees. The terrain I was walking through was defined not only by human divisions—New York and Pennsylvania, for example—but as part of a natural network of watersheds, interwoven ecosystems that connect communities across borders and boundaries.

The eastern woodlands sit atop piles of sediment that have been sculpted by running water since the Ice Age ended more than eleven thousand years ago. At Marie McRae's farm, for example, I had been in one watery world, as Six Mile Creek flows west, to Ithaca, where it drains into Cayuga Lake, part of the Lake Ontario Subbasin. Within my first few miles of splashing up and over Hammond Hill and descending toward Candor, though, all the water around me was flowing in a different direction: southward. This was part of what hydrologists call the Upper Susquehanna Subbasin, contours connected by veins of water across an area of five thousand square miles. All the water there flows into the Susquehanna River, which weaves not north to the Saint Lawrence, as the Six Mile does, but south through Pennsylvania to

Chesapeake Bay, where it delivers millions of gallons, more than any single source.

By late afternoon, as I approached Pennsylvania, I wondered at the consequences of drilling deep into such watersheds, penetrating layers of subterranean rock and water, bringing things intended and not to the surface. In the coming days I planned to follow the route of Apalachin Creek toward its source near a town called Friendsville, then move east across Susquehanna County to Dimock. Nearly a decade earlier some of the first fracking wells in the state were tested in Dimock, and more than five thousand gas wells had been dug across Pennsylvania since, stretching from Washington County, near Pittsburgh, to Susquehanna County, one of the most heavily fracked places in the state. Perhaps nowhere in the United States had been so quickly turned from nature to industry to feed the demand for more energy.

As I traced Apalachin Creek on a final mile south to the border, I passed on the east side of the road a football game hosted by the Apalachin Raiders: middle school kids suited up in the sun, cars parked on the grass, and families perched on lawn chairs along the lines. Joyous calls and barks of aggression rose into the air, but I was tired, my blister now making me think of every step. Soon I saw a rectangular blue sign with white lettering: "Welcome to Pennsylvania." Twenty yards beyond, more signs offered slogans: "Adopt a Highway Litter Control Next 2 Miles"; "Reduce. Reuse. Recycle"; and "Keep Pennsylvania Beautiful."

I ached my way around a bend, and the village of Little Meadows emerged in Norman Rockwell tidiness: mowed lawns and swept porches. The border settlement had taken root two centuries before, and one history of Susquehanna County written in 1873 defined its origins in relation to the watery world in which it grew: "Bear Swamp, not far from the head of the creek, is one of several marshes, almost amounting to lakes. . . . Little Meadows . . . is two and a half miles lower on the Apolacon Creek [Apalachin Creek], across which, at this point, the beavers once built a dam, and thus cut off much of the timber before it was visited by the ax."

Despite all the gas drilling elsewhere in Susquehanna County during the past few years, Little Meadows had remained an untouched outpost at the edge. But that was about to change. Landowners had joined together to sign the Friendsville Group Lease, a deal in which the gas company agreed to pay $5,500 an acre plus 20 percent of royalties to drill on more than forty thousand acres of land. As I walked into town late that Sunday afternoon, two drilling rigs rose high on surrounding ridges. Within weeks two more natural-gas wells would be fracked, and the residents of Little Meadows would have their long-awaited payday. One local had told me over the phone a week before I arrived: "You clearly are hitting this area at the right time."

※

The next morning I hopped into a four-wheel-drive buggy with Carl Ernst and rode off to see the action. Carl had been a chief negotiator of the Friendsville Lease, and he had offered to put me up at his property, Meadow Pond Farm, home to a fine house and riding school, when I passed through town. When I'd arrived the evening before, I dropped my pack and walked across steep pasture, past a bubbling brook, to a view of the surrounding hills. I could see the steel spikes of one drilling rig across the valley to the east and another atop a prominent ridge south of Carl's property. When completed, each well would descend more than a mile, with long forks of horizontal pipes fingering through the Marcellus shale beneath hundreds of acres of land. In the evening light the two rig towers looked to be little more than sticks emerging from the trees. But after dinner, with darkness fallen, the rigs east and south sparkled. They did not flare with the flames of natural gas, as their pounding down through the earth had not yet reached its mark. But they were lit in the night from electric lights, and those made the rigs shimmer in the sky, stationary satellites come to survey the terrain they were changing. A steady breeze delivered sounds of the pounding drills across the hills.

By day the towers had again been reduced to sticks among the trees, and Carl motored downhill into town. Carl moved with the certainty of someone already finished with a career in industry and enjoying rural order. He had long worked in research and development for International Paper, across the border in Binghamton, New York, before retiring to his hill above Little Meadows. When I'd asked him the evening before if he worried that the fracking fluid that would course into the layer of Marcellus shale beneath his horse pasture might cause damage to that bubbling brook, among other things, he did not seem worried. "Is there some risk? Yeah," he said, while feeding his horses, Jackie, Missy, and Chance. "I guess that, relatively speaking, I feel the risk is minor." Carl reminded me he was a chemist, and he was confident that if some damage were done, the gas companies would clean it up. He figured that in the big scheme of things, gas may as well be harvested far from city centers, where so many people enjoy the energy. "This is a pretty sparsely populated area," Carl said.

He preferred to talk about the terms that the group of landowners in the Friendsville Lease had negotiated for themselves. When the fracking boom had first come, unsuspecting property owners in other parts of Susquehanna County had leased land at $20 or $200 per acre, not the $5,500 per acre the Friendsville group got. And landowner rights to 20 percent of royalties, after expenses, offered promise of much more. In that way Carl saw the deal more for the financial impact it would have for his daughter than for the ecological impact on the land where he lives. "It's a generational asset that will span beyond my lifetime," Carl told me.

So Carl drove the four-wheel drive with an air of enthusiasm as we headed into town that morning, because for him, like so many of the 250 people living in Little Meadows, the drilling in the hills meant a personal profit soon to come. In town Sally Dewing, postmaster, stood behind her window and told a customer about the sight from her land, a few miles west, where other rigs had already done their work. "The flaring is fascinating," she said. "One time, we could see three different wells flaring at once." Then she turned to Carl and relayed a report from her husband about crews moving fast to bury more pipeline to ship the captured gas. "They had the brush all cleared, and the trees cut down," Sally said. "He said, 'They're really going to town.'"

On a steep pitch of dirt track heading toward the drilling rig east of town, Carl and I stopped to talk to a woman out walking. She marveled at the size and shape of the huge holding tanks for the water that would soon frack the well, which was named York, for the man on whose land the well was dug. "It's like a big built-in swimming pool," the woman said. And at White Tail's Bar & Grill, where Carl and I pulled in for a quick lunch, talk at the bar was all about the traffic bustling past on its way into the hills: tankers and dump trucks, graders and pickups. The bartender joked with customers about the names of companies on all the trucks. He suggested they each choose a favorite, as though they were football teams. "Which gas company are you rooting for?" he asked. "I'm rooting for Schlumberger. I just like the name!" But then he quieted as he described the sounds that came down from the drilling rigs in the dead of night. "It just amazes me," he said. "They keep me awake. I have to sleep with the TV on."  .

One customer—a traveler from a nearby town—countered that some aren't so lucky to have a place to rest, as rent in the area has more than doubled since all the workers came in. The bartender mentioned a check he'd received from the gas company. He found it curious that he'd already earned a royalty, as the gas hadn't yet begun to flow from the well being dug beneath his land. Carl pointed out that the check represented what is known as a "shut-in" payment, something to keep the lease current while the wells are dug. "Oh," said the bartender, with a turn of his head, "so that's just to lock me in?"

I'd already seen what the bartender and others were getting locked into, because before lunch Carl and I rode in that buggy up onto the ridges. We had sped west past Sally Dewing's place, which sells maple syrup in the spring, and Carl turned south as the road climbed toward the site of the Caprio well. The well, which would tap the gas beneath 640 acres, would reach into the Marcellus shale all the way beneath Carl's property, more than a half-mile north. But as the well was being dug, the impact was greatest along Abell Road, and there we stopped to talk with a man whose house, once hidden at the edge of a wooded cul-de-sac, hung above the clearing of the well pad, like a villa perched at the edge of the sea. The elderly owner opened his door, oxygen tank in tow, and gave Carl a report about the pace of progress. "Yesterday they started pulling pipe back out of the ground and stopped the engine," he said. He wasn't sure exactly what was wrong, but there had been some sort of safety concern on the site. We wandered around the side of his house to get a view at the operation. Only a few trees stood where once woods had been, and then the rig rose, surrounded by a cavalry of trailers and trucks.

Carl wanted to show me one more gas well, this one already drilled and readying for the hydraulic fracturing, which would send water, sand, and chemicals deep into the earth. The next ten minutes were a blur of dirt and gravel spitting beneath the ATV's wheels, thick forest walls opening only briefly before closing again, as we climbed one back road and then another onto a ridge farther south of town. The pad for this well, the Kropiewnicki, had been cut clean from a swath of undeveloped forest, with no houses or roads nearby. The stretch that we drove on the last approach had been graded just for this operation. After Carl made the last turn, the dirt road flattened on the ridge top, and a stout rubber pipe, easily two feet around, ran in a side ditch, leading the way toward the rig. In the coming days the pipe would deliver millions of gallons of water from an open-air tank to pump underground. The engine of Carl's four-wheel drive whined for a few hundred yards more, and then Carl took his foot off the gas, as the dirt road passed through a high chain-link fence and approached the edge of the pad.

A guard stood in a shack beside a deep red sign that read, "STOP." The sign was meant for the truckers hauling in tanks for the company, Talisman, but also for anyone else, curious, or concerned that maybe this idea of opening up the earth and shooting things down through layers of time in such a mad rush may not be the best idea. Each of the thousands of wells dug in Pennsylvania during the fracking boom, after all, was just another anchor committing society to a fossil-fuel future. As Carl and I idled in the buggy before that red sign issuing its command to stay put, the door to the shack opened. The guard emerged, hard hat on his head, arm extended with a spread hand of denial reinforcing my separation from the source.

⁂

Susquehanna County is shaped like a tidy rectangle and covers more than 800 square miles at the northeastern edge of Pennsylvania. The mountainous swath of terrain, set some 150 miles northwest of New York City and on a direct route between few places, sprawls as an East Coast equivalent of flyover country. Two days after my ride with Carl, I sat with my back flat against the trunk of a tree, one of two braced between the road and a field at the top of a steep, short hill. I had just spent ten minutes walking up the sharp pitch from the west, having left Little Meadows earlier in the day. My feet ached and I was thirsty, so I leaned against the tree with a rooftop view: ridges of the Endless Mountains spilling east and south.

A decade of fracking for natural gas in Pennsylvania before my arrival meant that a deeper industrialization was already digging in. Once the fuel has been captured, there is no stopping its march toward market. Already in Susquehanna County, after battles about contaminated water wells in zones of heavy fracking, the fight was moving toward keeping air relatively clean and guarding rural character in the landscape, as construction crews prepped more well pads, buried more pipe, and built more compressor stations to send the gas into an underground delivery network that stretched from Texas to Maine.

As I sat in the shade of the tree, I felt at once sheltered and exposed: the gas fracking that had swept across Susquehanna County and so much of Pennsylvania struck me not as technological success but a clear case of society's inertia carrying it in the wrong direction. Despite the dire warnings from scientists monitoring the increase of carbon in the atmosphere, the fossil-fuel industry was only expanding. The International Energy Agency projected that by 2050 global energy demand would increase by more than 60 percent, and larger amounts of oil, gas, and coal would still be driving the industrial economy. If that held true, the agency predicted, more severe climate scenarios could too, with average temperatures rising more than seven degrees Fahrenheit by 2100, accelerating sea-level rise and the destruction that comes with it. Yet all across Pennsylvania, politicians and their voters had welcomed the earliest prospectors who came looking to frack new wells for natural gas. And by the time I sat against the tree, more than a thousand wells had been dug in Susquehanna County alone, a staggering toll that represented one well for every thirty-nine people who called the county home.

I still had to walk more than twenty miles to reach Dimock, the rural hamlet that had been my destination from the start. There was no public land along my route, so that evening I would need the kindness of strangers if I hoped to find a place to pitch my tent. At the bottom of a steep hill, I saw a shuttered building that once had been an inn tucked in the valley. Wagons used to stop there so that horses could rest and drink water before climbing again. I barely broke stride, and soon I passed a dirt road that cut south. A sign announced that the road led to the Baker 1H and 4H wells, drilled by Carrizo Oil & Gas. Carrizo, according to the sign, had been granted a permit, valid from March 28, 2011, to March 28, 2016, to use as much as 2.1 million gallons of water per day to put the wells into action.

A bit farther along, on the north side of the road, manicured meadows surrounded a private estate that had its own airfield. The road quickly descended again, and I was in familiar terrain, a tight dirt lane passing through a hollow, with small cabins tucked among trees. It reminded me of the ravine behind my childhood home on the outskirts of Pittsburgh. I saw a man and a dog walking in an overgrown meadow, but they seemed like they did not want to be bothered. I turned up another road that cut between an aging farmhouse and its barn. I stepped onto the porch and knocked, but no answer. I moved on, and up around the bend I saw another house, and a man working on a window. He was wearing a Pittsburgh Steelers hat, and I made note of that when I called out, and soon Kevin Conway had offered me a cold Rolling Rock beer and a seat on a deck above a steep stream. The banks of the stream had washed out only days before, during the storm that had passed through when I was camped in Shindagin Hollow. Flooding had sunk many basements in a nearby town, but Kevin's house was spared.

He told me I could pitch my tent in his yard, and he began working the phone, calling his wife and then their grown daughter. He told his daughter to get her husband, stop at the store for chicken marinated in a locally famous sauce, and come over after work. There would be a barbecue on the deck. Then he went into the house and returned with an envelope from the gas company.

Kevin owns a small portion of the land accessed by a gas well a quarter mile away. The well, which harvests gas from more than 523 acres, includes 0.14 of an acre of Kevin's land. The letter informed Kevin that his portion of any profits, once divided by total acreage and multiplied by his 17 percent share of well royalties, would come to 0.00004544 percent, after expenses. So far, that equaled nothing. "Isn't that special," Kevin said, sarcasm thick in his voice. Kevin had signed the lease some years before because he saw the drilling as an inevitable force and didn't want to open the earth without earning anything. He wasn't holding out hope for any big payday. Kevin said of the gas companies: "They're going to lie to you or bullshit you."

He handed me another cold beer and a leather-bound volume published in 1873: *The History of Susquehanna County*, written by Emily Blackman. It was there on Kevin's deck that I first read Blackman's watery description of Little Meadows. She also wrote in fluid detail about my destination of Dimock. A settler from New England had built a log home in the township in 1802. "That dwelling stood in a beautiful valley, nearly surrounded by hills," Blackman wrote, "beside a brook of pure water which ran through, and gave name to the valley."

She described more broadly the lay of the land in Dimock, as it was founded in 1832, the nineteenth township in the area. "Excepting a slight alteration of the line between Dimock and Jessup, its dimensions have remained as at first, six and a half miles east and west, by four and a half miles north and south. From the timber frequently found here it has been sometimes called 'the Basswood township.' With the exception of the outlet at Elk Lake and near tributaries, the township is wholly drained by the Meshoppen, or Mawshapi, in Indian language, signifying cord or reed stream."

⁂

Sixteen miles and two days farther along, I drew to a stop, my walking stick as a leaning post, along the Meshoppen Creek Road. I stared at the creek as it meandered in hairpin turns through a meadow. A stiff breeze rocked the summer green and whipped the surface of a small pond, thick with lily pads. The ridges that framed the valley were gentler than out in the far west, but the view reminded me of one I'd seen years before in a remote reach of Yellowstone National Park.

In 2007 this oasis along Meshoppen Creek, and on the other side of Ely Mountain, had become an early demonstration of fracking's deeper damage. Cabot Oil & Gas, a publicly traded company with roots in Texas, had come into Pennsylvania, and of all the hills and valleys across the state, and even across Susquehanna County, had chosen a particular ridge around the bend from where I stood. There they drilled an exploratory well six thousand feet down to the Marcellus shale. That showed so much promise that soon the secret was unlocked, and neighbors all around began signing leases for little money, some for only twenty-five dollars an acre. "Everybody was kind of assuming they'd give you the money and never come back," one resident would tell me.

But Cabot did come back, putting a well up on Ely Mountain, just to the west of where I stood, and others throughout the hills. All told, more than sixty wells were dug in the Meshoppen Creek watershed. On one stretch of road north of Ely Mountain, more than a dozen water wells became contaminated with high amounts of methane. Several residents filed lawsuits against Cabot, which denied any role in the problem. Some suits settled quickly, others dragged on for months or more. In 2010 a consent order from the Pennsylvania Department of Environmental Protection found that eighteen wells had been contaminated as a result of the Cabot drilling. In Dimock neighbor became pitted against neighbor, as the rural township's to-drill-or-not-to-drill debate hit the national stage. The filmmaker Josh Fox featured the damage done in *Gasland*, his documentary exploring the dangers of fracking across the country, and politicians, celebrities, and news crews descended.

All was quiet, four years later, as I followed Meshoppen Creek south around Ely Mountain. I passed a steep rocky side road, and two stout signs advertised the Ely, K. PAD 1 and Ely, K. PAD 2. A half-mile farther, the sound of intermittent gunfire—someone shooting a .22-caliber rifle in what must have been practice or play—came through the trees from the east. The report of each shot took me back in time, as though a settler dressed in buckskin and furs were navigating the still-virgin forest, on the hunt for rabbits or something else for that evening's dinner pot. But the bounty now was deep underground, and on the other side of Ely Mountain Victoria Switzer measured her days in the bottled water delivered to her door.

Victoria and her husband, Jimmy, had leased gas rights to Cabot early on. After her well began to turn up too much methane, she became among the most vocal and visible of the critics of the drilling. I sat with her for two hours, resting at the long kitchen counter in the timber-frame house Jimmy and she had built. She recounted some of the people who'd come through Dimock: Robert Kennedy Jr., Susan Sarandon, Yoko Ono, and John Hanger, the head of the Pennsylvania Department of Environmental Protection. "We were the circus," Victoria told me. "We were the freak show."

Cabot countered early claims by saying that water around Dimock had never been very good. Victoria and other neighbors, several of whom live along Carter Road, just north of Victoria's house, and who settled lawsuits against Cabot, still have drinking water shipped to their homes. "I go to bed at night, and I don't own this property," Victoria said of the gas lease that controls her land. "We're just a tenant now, and we have a very nasty landlord."

By the time Victoria and I talked, there was little debate that fracking can damage ground water. To prove the point Victoria opened her laptop and called up a story published the morning of my visit by the *Pittsburgh Post-Gazette*, which had done deep investigations of the impact of drilling across the state. The state Department of Environmental Protection had been harshly criticized for years for doing little to regulate the gas companies. But, according to the *Post-Gazette* report that Victoria showed me, the DEP had determined that the drinking water supply at three residences in another county, Westmoreland, had been damaged by another gas company, WPX Appalachia. The homes were located more than three hundred miles from Dimock, many watersheds away. This news foreshadowed another report months later that documented roughly 240 private water supplies statewide had been contaminated by fracking operations during a seven-year period. "Dimock," Victoria told me that morning in her home, "is not an anomaly."

She soon expanded our conversation to the topic that consumed her most: the increased industrialization that has come to the countryside since the early days of drilling. She had recently signed on with another woman to lead a group fighting for clean air in Susquehanna County, as the pipelines and compressor stations that came in after the gas wells were dug create pollution of their own. "Everyone wants to see this American energy face," Victoria said. "But behind it is a reptile."

Victoria was talking about more than herself and her neighbors living in the shadow of Ely Mountain when she said, "Where are we? We're still under the thumb of the fossil-fuel industry." And she feared what the future would bring to places across the United States and around the world, where the fossil-fuel harvest continues to accelerate. "It's not only about the fracking," she said. "It's about all the accidents and permutations that occur when you industrialize a nonindustrial area."

It was a balmy afternoon when I arrived at the Teel farm, just south of Victoria's home. Ron Teel, born on the farm and returned after years of truck driving around the United States, sped up to me on an ATV but stopped only briefly. He told me to haul my pack up the hill, where I'd meet his wife, Anne, and then he raced off to tend to a horse. It was a good five minutes straight up and across an open field to a brick house, and before I knocked on the door I looked back to see the land fall away behind me, a late summer Eden of grass and green forest. The farm covered hundreds of acres of hills and valley carved to the south by the still-winding course of Meshoppen Creek. I could see below a red barn and several acres of hay growing high.

Anne opened the door and, once convinced that Ron had sent me, invited me warmly into a country kitchen with a suburban tidiness. She offered a cold drink, and for more than an hour we talked about the Teels' enthusiasm when Cabot first came searching for gas eight years earlier. The Teels had figured if any drilling were to happen at all, it would be limited but offer a chance for some income to offset the costs of keeping the land. So they signed a paper that, Anne said, contained one paragraph that did not indicate in any specific way that less than a decade later the Teel farm would be home to eight working wells and more than four miles of underground pipeline.

Those wells had paid the Teels handsome profits, and for years the Teels had been local champions, speaking out in the media about the good things drilling would bring, even as they and neighbors saw their wells contaminated. But as we sat and sipped and waited for Ron to return from the barn below, Anne chronicled their hard-earned change of heart. "There were a lot of lessons to learn, and unfortunately we were the guinea pigs," Anne told me. "We had no idea."

That original lease mentioned terms such as *ingress* and *egress*, and on the Teels' land those terms translated to a compressor station the size of a football field. It was set in the valley below the house, not far from Meshoppen Creek, and it roared to life twenty-four hours a day, gathering gas from nearby wells and shipping it off to New York City and beyond. "The noise," Anne said. "The blow-offs and squeals. And then you wonder what kind of chemicals are flaring out of there."

In the months before my visit, the Teels had been suing the company running the compressor station, as the company had proposed burying six thousand more feet of pipeline underground. The plan was for that pipeline to traverse the only remaining acres of untouched land on the Teel farm. "We already have twenty-two thousand feet of pipeline on us," Anne said. "We don't want anymore."

Ron had come back to the house, and he suggested I climb into his old pickup truck to take a tour of the land. His parents still lived in a house below their big hill, but we didn't stop there, angling instead on a dirt road along Meshoppen Creek toward the compressor station. "We didn't know 'Marcellus.' What's that? Never heard that word in my life," Ron said, recalling his reaction when the first land man turned up at the farm. "Gas? Here? What are you talking about?" He recalled with dark meaning the words of a young engineer sent out to work on the first well in the summer of 2007. Ron did not pay much notice to the words back then, when the engineer stood on a simple pad in the middle of a generational farm and said, "You're going to see huge changes here in the future."

On a dirt road running from the fields into forest, Ron pulled to a stop in front of a chain-link fence, as though approaching a bandit's hideout. He unlocked the gate. "Everyone's got a key to this place," he said of the crews, coming and going. "Any sense of privacy is totally gone." The day before I arrived, Ron and Anne learned that a Pennsylvania Superior Court panel had ruled two to one against their attempt to deny more pipeline access to Williams Field Services, which had purchased Cabot's rights in 2010. But Ron was determined to appeal. "It's become a huge part of our life," Ron told me.

We climbed back into the truck, and Ron followed a series of single-track turns that he'd navigated since his teens. The road climbed toward a pasture atop a high hill. Ron pointed to where the Tennessee Gas Pipeline, one of the longest single gas-transmission trunks in the nation, passes through his land on its journey toward Manhattan. The old truck dug in harder, as the road pitched steep and the land all around fell away behind us. Ron was navigating toward the site of Teel No. 1, the first exploratory well that Cabot had drilled seven years before, when the state and its residents saw only the promise of unexpected profits. "These guys came in here and did as they wanted," Ron told me. "There were no inspectors. There was nothing."

It was late afternoon, and the air softened from high humidity toward the ease I remember from childhood summer evenings. Shadows had grown long, greens richer all around. Ron made a final few turns toward the well. We would get out together, Ron standing with his hands in his pockets and staring a dead stare at the simple contraption: a five-foot-high stack of pipe and steel parts, all a bit rusted, continuing to draw gas from thousands of feet beneath the surface. "When is enough enough?" Ron said. "You can't take and ruin someone's property. There has to be a limit. What is reasonable?"

※

That night, after Ron and Anne had long since gone to sleep indoors, I lay in my tent beside the Teels' brick house. The compressor station roared at odd intervals in the forest below, as more natural gas flowed in the Tennessee Gas Pipeline, beneath the farm and Meshoppen Creek and on toward New York and New England. Did that natural gas make it all the way to my house in New Hampshire, where my wife and children, at that very moment, slept? Surely some of what heats our home came from the wells in Susquehanna County. The sense of dislocation I had experienced at home was shattered with the steps I'd taken to Dimock. My family's consumption was, in that moment and always, intimately connected to the ridge on the Teels' farm, where I lay in darkness above the trees.

I would soon return home to New Hampshire. But I planned next to confront what seemed a final terrain of fossil fuel: the coalfields of Wyoming's Powder River Basin. I hoped I would find there not an industry thriving, but weakening toward its end. That, perhaps, would open a chance for renewable, noncarbon fuels to flourish more quickly, offering a more sustainable system, better balanced between nature and industry.

Sometime after midnight on the Teel farm, I drifted to sleep, but woke at first light. I looked from my tent east across the valley. All was quiet and shrouded in fog, as though a scene from a battlefield painting, fragile peace between bouts of violence. Then, from below, came the grind of a motor and the squeal of a flare and the vibration of more gas compressed for transport to distant markets. Its capture echoed out of the valley.

# Chapter 3
# Black Thunder Bust

The train and its hundred loaded coal cars marched around the bend and away, into the rolling prairie of eastern Wyoming. I heaved my backpack over the barbed wire. It hit the ground with a thud. I tossed my driftwood walking stick next, and it bounced with the play of a pebble. I swung myself up and over the fence, landing on the other side on all fours. Sharp stones pressed into my palms. The air smelled of dirt and diesel.

Across the tracks and atop a steep, short hill ran two paved lanes. I knew that to the east the road intersected gaping canyons of excavated earth, among the largest open-pit coal mines in the world. The craters belonged to the North Antelope Rochelle and Black Thunder Mines. Those belonged to Peabody Energy and Arch Coal. The two mines produce more coal than any others in the United States, and their canyons cover an area larger than metropolitan Washington, DC, yet they were home to no one. The earth that lay before me had been so taken for the harvest that it long ago was dubbed a "sacrifice zone."

I slung my pack onto my back and stepped quickly across the four sets of iron rails, then slogged up the hill to County Road 37. Silence. I turned east and walked slowly. The road crossed above more train tracks that angled north. I stared down from the bridge into idle, empty coal cars. Each yawned up at me, at rest but ready. A local rancher, L. J. Turner, had driven me into the grasslands eight miles south of the mines and dropped me off early that morning, and I was thirsty from a first day of full exposure, walking steadily across rugged terrain. I stalled, unsure. Swallows flew from their mud nests beneath the bridge. The black beacons swooped and squeaked overhead, calling out questions about what I, a lone human, was doing on foot in this land of relentless machines.

Wyoming's Powder River Basin, which sprawls across the northeastern corner of the state, is roughly the size of Massachusetts, but home to only 1 percent of its population. Only seventy-one thousand people live in three large counties that cover thousands of square miles, and most of those people cluster in the city of Gillette, fifty miles north of where I stood.

North Dakota's oil field had been a dizzying introduction to a fuel landscape in a moment of creation. In New York and Pennsylvania I had traversed extremes, from one place where people decided not to unearth more fossil fuel to another redefined by the industrialization that follows when they do. In the Powder River Basin, which for more than a decade had provided nearly half the coal in the country, I would face the depths of fossil-fuel dependence: a landscape completely devoured to feed the United States' appetite for energy.

During the week ahead I hoped to walk forty miles among the mines and out the other side, moving along tight roads and sleeping at night on small stamps of federal land still undisturbed on the surface. My meandering route north, then east, then north and west would form the shape of a giant question mark, as I met so much of what the industrial world demands but the planet can no longer endure. As I stood on the bridge, swallows continued their chorus above me. I had spent hours that first day traversing open grasslands, following the tracks of antelope and prairie cottontail through shallow draws and across treeless pasture. The North Antelope Rochelle Mine lay just two miles away. I needed rest before I would be ready.

A few hundred yards beyond the bridge, I cut south into a flat piece of land bound on one side by the asphalt lanes, on the other by the four sets of train tracks I had already crossed. There are few trees anywhere in eastern Wyoming, and the tallest things along the roadside were waist-high sagebrush bushes that framed a serpentine creek. The ground was hard-packed and nearly grassless, but mounds of earth rose around holes dug by black-tailed prairie dogs, who stood on hind legs and barked in high-pitched alarm as I walked past. I slumped to the ground behind a bush, just above a steep bank that descended to slack, brown water.

On the ground all around me, particularly near the roots of each sagebrush bush, little holes an inch or two wide led into burrows that bumped up trails of earth. I suspected that voles lived inside. I thought of rattlesnakes, as the temperature that day had climbed well above eighty degrees. But my stillness felt like safety. Meadowlarks and black buntings sang. The sagebrush shielded me from the road, and I seemed of little concern to the Northern Harrier that circled above the creek. A cottontail picked up my scent and scurried behind a bush, but that was all.

The valley lingered like land at the edge of a war zone, a buffer long since cleared of humans. I found unexpected peace in my exhaustion, a sense of strength in my solitude. I was glad to return to the demands of simple effort—the "right work," as Marie

McRae had put it in New York—that I had come to enjoy during my walks in oil and gas country. My shoulders ached. My feet itched with hot spots. My body and mind were alert, and I sat in soft breeze that afternoon feeling a giddy joy, certain that my steps were taking me someplace profound. I had been overwhelmed by the scale of the oil harvest in North Dakota. I gained strength of spirit in the debate about the wisdom of unearthing evermore in the gas country of New York and Pennsylvania. I hoped in Wyoming to encounter an end to the coal harvest in the days ahead. I knew that I would need more calories after that long first day walking, but my emptiness had not yet turned to hunger.

After an hour I pulled out my camp stove and dug through my pack for a canister of propane and a serving of freeze-dried chili. My home for the night depended on fuel from afar. I had set out carrying two gallons of water, and I still had more than a gallon left. I poured some into a small aluminum pot. I lit the stove and waited for the water to boil.

I heard the train and felt its pulse before I saw it, and then it was there, pulling to a stop on the tracks just across the creek from my perch behind the sagebrush bush. Two towering diesel engines stood before more than a hundred empty coal cars that trailed around a bend to the west. The train was too big for the valley. The engines, now standing just two hundred yards from me, shuddered electric sighs. I could not see anyone through the high windows. Meadowlarks and buntings still sang.

There is no discreet way to ship coal, no pipeline buried underground, no closed tank pulled behind a truck along interstates. Moving crushed rock at great volume takes brute force, so each day more than seventy-five trains roll from the Powder River Basin coal mines into the United States. Each departing train carries fifteen thousand tons of rock to power plants in Washington, New York, Minnesota, Texas, and elsewhere. The typical U.S. consumer burns the energy equivalent of twenty pounds of Powder

River Basin coal each day. One train of coal can fuel the lives of four thousand or so people for a year. But that is just a figurative reckoning, as each load is burned immediately. The shuttling of trains from mines to distant plants continues around the clock. Bottlenecks in the delivery happen only briefly, as empty trains returning to the mines have to wait for a loading train to clear out of the way.

I realized I had made camp for the night opposite just such a staging spot. The empty cars stood obediently as the engines ached to go forward. I decided to pitch my tent on a patch of matted grass at the creek's edge. It looked like it had been a bed for antelope or mule deer not too many nights before. I snapped the aluminum tent poles together, then thread the nylon fabric along them. The temperature would not drop far, but I put on the tent fly, if only to provide a sense of separation. At twilight several coyote pups yipped and yelped on the opposite bank of the creek, between the train tracks and me. I wondered about mountain lions, which still roam. Before dark I climbed into my tent and slept.

At four in the morning I was alert again, eyes open wide, shoulders and legs startled against the ground. Another set of idling diesel engines just across the creek growled toward movement. Wheels squealed on steel rails. The coal cars clicked together one after another, the crashing of couplings cascading the length of the train and back, all in the valley shuddering beneath the sound. Coal emits more carbon dioxide when burned than any other fossil fuel, and it had been an early target of efforts to curb global warming, as market forces and government regulation made it more expensive to harvest and burn. But in that summer of 2015, coal still powered one-third of all electrical generation in the United States. The train opposite my campsite accelerated, undaunted, to deliver another load, confident that when it reached its destination, consumers, so reliant on the comfort coal brings, would cheer its arrival, or not even know it was there.

*

Before it fueled the birth of the Industrial Revolution, coal had been scratched from the surface for thousands of years: the Chinese in 3500 BCE, Romans in what is now England, Hopis in the U.S. West four hundred years before the United States was formed. But such tinkering with the black rock did relatively little to change how humans lived. Then, less than three centuries ago, coal began to fire furnaces that steamed ships and forged steel, and everything began to move faster and climb higher. As early as 1835 Frenchman Alexis de Tocqueville was overwhelmed by burning coal while wandering the streets of Manchester, England. In an account of his journey, he described the sun, that distant star whose energy had created the fuel, as so shrouded in smog that it looked like a "disc without rays."

On the U.S. side of the Atlantic, much early mining to feed growth took place in Appalachia, particularly in eastern Pennsylvania, where deep seams held anthracite coal that burned hot. There, and in many other places, the coal lay so far underground that it was simplest to send people into the earth. Tunnels dug straight down, splintered in subterranean trails. Picks and pans in hand, miners crawled deep, extending the network by removing dirt and rock and setting wooden beams to brace the world overhead. Such scrambling underground in search of more coal still happens around the world—in the eastern United States, India, Russia, Australia, and elsewhere.

Fifteen years before I walked beneath Wyoming's big sky, on a bitter February day half a world away, I followed a man named Nikolai Galushko onto a swaying iron elevator, then turned to see a woman named Galina swing shut the heavy gate with a shrug. Nikolai and I were surrounded by other men beginning a shift at the Oktyabrskaya coal mine in eastern Ukraine, and as the elevator descended, groundwater flowed from the walls. During five minutes we traveled more than a half mile underground—as deep as the world's tallest skyscraper climbs high. Then the elevator slammed to a stop, and the dark of the depths was erased only after we switched on our headlamps. We stepped into a tunnel that measured six feet high and wide.

I had gone to Ukraine as a journalist to write about the mines, where frequent methane explosions killed dozens of miners. Ancestors on my mother's side of the family had migrated in the early 1900s from Welsh mining country to eastern Pennsylvania coal towns, and I wanted to feel the emotions of that life. So I followed the men into ever-narrowing tunnels. The ceiling dripped constantly, and the air smelled empty as death.

After walking nearly a mile underground, we descended a steep slope, and Nikolai and another man dropped to their bellies and soldier-crawled into a three-foot-high opening that led to the coal seam. As Nikolai looked back at me, I could see his eyes gleaming from a face caked in coal dust, and there was something more than desperation. I dropped to my stomach and inched toward him. A man named Gennady Iskandiar slid past me, his lamp flashing toward the coal seam. He carried a pick with a wooden handle. The exposed coal lay more than a hundred feet in front of us. I made it only ten or fifteen feet. As I rose to my knees to shift my weight, my back scraped against the ceiling. It seemed as though the earth above and below were closing in. I yelled out to Nikolai that I did not trust myself to go farther. And I lay there, my face pressed against the only rock I could see in the arc of my lamp.

My descent into the earth with Nikolai, Gennady, and the others was a stark reminder of the extremes humans go to underground to capture what society demands above. At the time of my visit to Ukraine, in the winter of 2000, eastern Ukraine was struggling through a decade of transition after the collapse of the Soviet Union. State services were sketchy at best, corruption was widespread, and individuals were left to fend for themselves.

Once I'd ridden the elevator back to the surface, I navigated cold hallways and slept in a room poorly heated, with windows that did little to keep out the cold of the February night. On the edge of the city of Donetsk, small piles of crushed coal littered a loading yard. There I saw a lone man scrambling, sack in hand. He could grab enough coal in one load to warm his home for a day.

※

In the open-pit mines of Wyoming's Powder River Basin, there is little intimacy in the transaction, as the surface is completely removed to expose the coal. On my second day of walking, less than an hour after I left my camp by the train tracks, I reached the North Antelope Rochelle Mine, and I spun at the edge of County Road 37, daylight exposing everything. A dump truck, appearing as small as a toy in the open terrain but with tires more than two stories high, backed up to a flat ridge just east of the road. Black smoke belched from double stacks as the truck raised its bed. The steel bucket lifted slowly, and dirt and rock avalanched down a steep slope. There the dirt and rock settled, leaving the slope a bit larger, a makeshift mountain alive and expanding. Up and down the fifty-mile band of mines, and at other mines north of Gillette, such holes are ever evolving, their outer limits shifting over the course of weeks, months, and years. As earth is dug at one end of a hole, more dirt is backfilled at the other, so the canyons actually move, roaming across the prairie like open-mouthed amoebas.

The dump truck, its bed lowered back down, drove away and into a canyon beyond. Four or five minutes passed, but not more, and then a second truck emerged from below, backing up to the ridge. That truck's bed rose just as surely as the first, and more dirt and rock rained down. A cloud of dust kicked up beneath the belching black of the truck's exhaust, and the slope grew again. The land all around was gray and brown and brittle as a desert in drought, and I could feel the origins of this manufactured landscape: smoothed slopes and angled ridges, piled rocks and potted holes.

Coal in the Powder River Basin lays in luscious seams a hundred feet thick, and those are buried just beneath the surface of the prairie. The coal seams began to take shape as much as 130 million years ago, as buried plants turned to peat and eventually to black rock holding energy that needed only to be unburied and burned. The Wyoming coal stayed beneath the surface until the 1800s, when settlers moving westward gave up hopes of more fertile terrain and stopped on the arid grasslands. A few began to dig in the dirt, and homesteaders would stop by with wagons to pick up fuel for their stoves. For a century and more mining largely remained a local endeavor. Then, with the passage of the federal Clean Air Act in 1970, and amendments in 1990, which required power plants to emit less sulfur, states around the country looked toward Wyoming. Powder River coal does not burn as hot as coal found elsewhere, in the eastern United States, for example. But it does have lower sulfur content. And in recent decades Powder River mines have produced roughly 40 percent of the coal in the country.

I set my backpack on the shoulder of County Road 37 and watched as the second dump truck worked the ridge. The truck lowered its bucket and drove back into the open pit, disappearing toward the dragline, a huge crane with a scoop that waited with more dirt and rock. The pit had been blasted away during months of effort, explosives loosening the earth, layer by layer, until a series of plateaus—*benches*, they're called in the mines—stepped down the side of the canyon toward the coal seam. That morning the dragline continued toward its target, lifting scoop after scoop of rock and dirt for the shuttle of trucks. The pit was just one small pocket within the North Antelope Rochelle Mine, and elsewhere crews were working on seams of coal already exposed, delivering the coal to crushing machines that shattered the rock into pieces

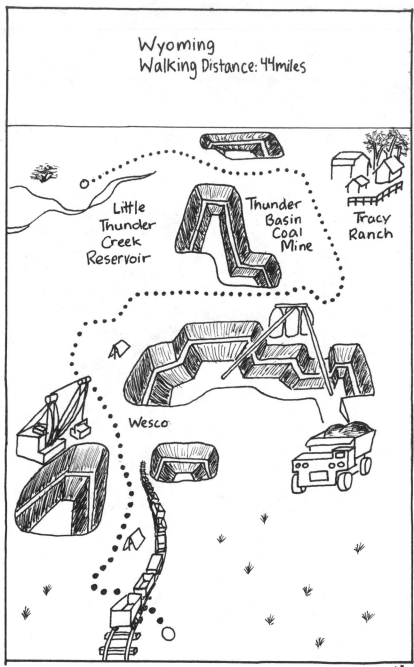

as small as a fist. The coal was then carried on conveyors to towering silos that stood over the tracks, where empty trains would arrive for the loadout before rumbling into the United States.

County Road 37 was quiet at midmorning as I sat on a small
rise and stared at the dump trucks. I spotted no people anywhere,
not even in the trucks. But I knew they were there, sitting in air-
conditioned comfort as they pushed pedals and turned wheels
that steered the machines. Even at the source there was a dis-
connect between nature and the people harvesting from it. In
these open-pit mines most risk had long been removed, and there
was no question of control. I kicked at gravel as the morning air
gained heat. A sign nearby warned, "Blasting area. Blasting emissions possible. Avoid contact."

Before I had arrived in Wyoming, L. J. Turner, the rancher, told
me I could camp in a pasture he leased from the federal government. That pasture lay eight miles north of where I watched the
dump trucks work. But after walking only a few hundred yards
up County Road 37, I heard the grinding of more machinery, this
time to the west. I cut across the road and climbed a low hill until
I came to a barbed-wire fence that marked the east edge of the
Antelope Mine. As I had been plotting my walk from afar during
the months before, I'd largely ignored the existence of this third
mine, set south and west of the Black Thunder and the North Antelope Rochelle Mines. It is one of the oldest in the basin, but
small by today's standards. On one map I consulted, the drawing of a pick and shovel indicated its existence. But the Antelope
is large enough in real life to consume a small city, and from my
perch atop the low hill the earth dropped several hundred feet
into a deep canyon at least a half mile across.

At the far side of the pit from which coal had already been
claimed, a backhoe was dumping tons of dirt down a slope, and
below three giant bulldozers worked in synchronized precision
to fill the chasm. Each time a fresh pile was placed, a bulldozer

would push it in a straight line the length of several football fields, leaving a trail that added corduroy contours to the land. I watched as one bulldozer neared the end of its path, where the ravine dropped deeper still. Then the machine did what must have been a perilous dance, pushing the dirt over the edge, its own plow hanging into the air for a moment before the traction belts slammed into reverse and the bulldozer scurried backward to safety.

The Powder River Basin is a testing ground for the most innovative earth-moving machines, each larger and more powerful than the model made before. Along the highways entering Gillette, equipment yards look like munitions depots, with giant cranes, dump trucks, graders, dozers, and more shining in anticipation of being sent out into the mines. The dump trucks alone could tell the story. In 1970 a truck held less than fifty tons. Today the largest carry nearly five hundred tons.

The Antelope Mine is owned by Cloud Peak Energy, which, like other corporations that lease the land from the federal government for private profit, promised to repair the place when it was done. But after decades of mining up and down the basin, less than half of the 170,000 acres dug up for coal have been returned to any semblance of prairie, and most of those only to the early stage of smooth dirt, piles of rocks for critters, and grass seed thrown to take root, which it only sometimes does.

When I had followed County Road 37 across the wide run of Antelope Creek earlier that morning, what once had been a steady stream was bone dry. I watched the bulldozers work to overcome the vacuum of open terrain: millions of pounds had been picked up and pushed around, everything in the way rerouted or removed. Even if they put every rock back in place, would there be any natural design left at all?

I was sure I could stand and stare at the bulldozers all day and see the same routine, the redefinition of a place built during millions of years. But I worried about my stamina beneath the heat, and I needed to cover distance. I turned back to the road, which soon arced to the west, and, beyond that, I found myself alone with an antelope. He had a small rack atop his head, and he twitched his white tail as he looked my way. County Road 37 was bound by barbed-wire fence along each side, and I wondered if the antelope was trapped between the fences, like me. He waited at first as I took a few steps toward him, then he bolted farther up the road. He stopped after a few hundred feet and grazed on thick grass on the roadside. As I got closer, he stared, then bolted again. Did he wonder if I might be a mountain lion? Our stop-and-go dance continued for more than a mile, and together we traveled past the north edge of the canyon of the Antelope Mine. Finally, the antelope leaped over the barbed wire, with a jump that could not have taken much effort, onto a piece of smooth earth.

An antelope's gait is at first glance graceful, fore and hind legs pinwheeling as the body stays steady above the ground. But the pinwheeling quickly appears to be on the verge of disorder, and this antelope looked awkward out there, bounding along first one way, then another. He stopped and lowered his head to eat but did not keep it down long. I sat on a gravel patch by the road and ate an orange that had been beaten around inside my pack. I had one bottle of water for the final miles to where I hoped to camp, but I had no idea if that was enough.

By midafternoon, the heat punishing, I began to lose focus on my steps, each no different than the one before. Suddenly, and as a total surprise, a scatter of buildings appeared on the right side of the road, looking like some kind of outpost on the surface of the moon. A sign in front announced, "Wesco Ground Breaking Service." The prospect of unexpected company grabbed me. I stumbled into the dirt lot. I dropped my pack outside a trailer that served as the main office and knocked on the door, then turned the wobbly handle. Inside there was no one. A small desk stood on the right. An air-conditioning unit hung in a window beneath flimsy shades. Dust everywhere.

I called out, "Hello?" No answer.

Louder then: "Hello?" Silence again.

I stepped back into the sunlight. I had been rationing my water all morning, and I was surprised at my sudden need: I was desperate to rest in shade and enjoy the cool water that must be somewhere inside. Across the lot, all hardscrabble and ruts, a high aluminum shed held a machine shop. I hustled that way. I stepped through a door that led into a break room. Cracked plastic chairs lined the walls, and there in the corner, a five-gallon bottle promised "Wyoming Quality Water."

I called out, nervous and impatient: "Hello?"

Nothing.

I hurried back outside and headed toward the corner of the shed, where a high garage door stood open. A truck parked in front sat empty, but its radio crackled, a dispatcher somewhere calling out orders. Rock and roll blared from speakers inside the bay. I walked to the door.

"Hello?!"

Had everyone been called down together into a mine to set a blast for some serious groundbreaking? I stood before the open bay door and tried again: "Hello?!" My voice struck me as strange, at once loud and lost in an echo of wind. I thought again of that break room, the water dispenser, its power cord keeping the cooler humming. Should I just take some?

After a morning alone in terrain stripped of rivers and streams and so much more, it seemed to be a test: who owned the water in that cooler anyway? Had so much earth not been moved, cutting the natural flow of water, I could have filled my bottles that morning in Antelope Creek. It seemed that a few quarts in return was only fair. I ran to my pack and grabbed two empty bottles, then doubled back to the break room. I burst through the door.

"Hello?!"

My fingers fumbled as I undid the caps, and I gulped down the warm water that had been left in one bottle. My heart raced as I held back the tap of the cooler, and I rocked on my feet as first one bottle filled, and the second. My pack was heavier, then, as I hustled back onto County Road 37. But I stopped short above a silver wrench that lay on the pavement. Had it fallen from a truck heading off to set explosives? I stared down at the single tool, forgotten and at rest.

The earth does not always move willingly. And when machines are not enough to do the job, explosives packed with chemicals are pushed down into the pit to send the very bottom deeper still. The goal is to blast away *overburden*, the rock and dirt covering the coal. Among the many things that rise up into the air after such an explosion is nitrogen dioxide, a toxic chemical that is regulated in most places where it can come into contact with people. Get a little dose, and eyes itch and mucous runs. Breathe it too deeply, and then can come headache, fever, chills, vomiting, and a collapse of the respiratory system: death.

Above the coal pits of the Powder River Basin, floating clouds are common after a blast, and people have learned to live with them. Sometimes a cloud will linger especially long, and then questions come. A year before my walk one thin blanket hung high above Gillette for more than an hour. Residents in the communities below were warned to stay indoors. But a county official tried to calm concern among those that make their living in the mines. "It's not poisonous," he told the local newspaper. "It's not toxic, but it could have long-term health effects with repeated exposure. I'm not trying to downplay it, but I don't want to give people a heart attack."

There is little choice for cover deeper in the sacrifice zone, where I walked with my freshly filled bottles of cold water. Environmental groups have petitioned the federal government to enforce existing laws limiting the chemical blasts, but down at ground level there are only signs and sounds of warning.

In late afternoon, as I continued toward the pasture that L.J. leased near the largest canyon of the North Antelope Rochelle Mine, I heard from the depths the call of a siren. The tinny alarm at first seemed comical, as though a toy. I knew that such an alarm rang ten minutes before a blast. Another alarm, lazy but urgent, like a British police siren, sounded one minute before the earth moved. Then: a roar of low thunder, and a bulbous cloud—brown at first and tinged with orange—rose from the canyon. The airborne debris grew from the center in every direction, stacking on itself like one of the puffy white clouds above it. For one minute, then two, the brown cloud rising began to shift to orange, as dirt settled and the chemical creation drifted upward.

L.J. and his wife, Karen, had warned me to stay away from such blasts. But I realized as I stood and watched the cloud rising a half mile to the east that I knew little of how nitrogen dioxide spread, or settled. The orange road signs that warned to "avoid contact" were meant for people in trucks, who could turn and speed away. Karen told me that many calves they pastured near the mine died each season. She suggested I cover my mouth if I should be too near a blast. But would that really help?

I pulled the top of my T-shirt over my nose and held it there, but not for long, as I noticed the wind was carrying the cloud away from me. As it stretched hundreds of feet into the sky, the chemical cloud's edges expanded like the canyon below, shifting in a succession of shapes, one of which seemed, for a moment, to be a person bent and digging with a shovel. That soon dissolved into something else, like sagebrush blowing about, and the cloud separated from the canyon completely. The toxic plume shone orange as it drifted amid white clouds. Like the antelope pinwheeling across the newly planted soil, the plume looked unsure of its place in this manufactured terrain.

Before I left home to walk in Wyoming, my wife had given me a book called *Narrow Road to the Interior*, by the Japanese poet Matsuo Bashō, who had gone off wandering on foot a number of times, most notably when he headed north from his home village of Ueno in the spring and summer of 1689. Bashō's walk with a friend across the interior of Japan's main island, Honshu, was a quest of encounter, as earlier literary success had surrounded him with a throng of adoring patrons and students. By walking Bashō wanted to go beyond those distractions, to reconnect to, among other things, the Zen concept of "original mind." As Sam Hamill, who translated the volume of *Narrow Road* my wife had given me, explained in the introduction to the book, Bashō "longed to find his own deepest personal connection to the very real world that lay only far beyond."

As Bashō set out from Ueno, he posted three lines of verse by his door:

> Even this grass hut
> May be transformed
> Into a doll's house

I had tucked my copy of *Narrow Road* near the bottom of my pack, alongside a few notebooks and pens, and I didn't think to read it until I woke in the pasture the morning after the blast. The pasture, which was owned by the U.S. Forest Service, sat alongside County Road 37 and a northward extension of train tracks heading toward the Black Thunder Mine. L.J. moved cattle to the pasture to graze each summer, but he hadn't done that yet, so I had the place to myself.

I had pitched my tent amid stone-dry cow patties a few hundred yards east of the tracks and the road. As first light spun across the dew-drenched grass, I hung the tent fly over the fence of a cattle pen to dry. Daylight again exposed the coal-mine canyon, painting some rock walls with quicksilver, casting others in cold, dark shadow. Shifting color seemed to mark time that passed slowly, yet right before my eyes.

I was not ready to take the first steps that would lead to another day of movement, so I crawled back into my tent, lit my cook stove, and set water on for tea. Then I picked up the copy of *Narrow Road*. The slim volume was three inches wide and three and a half inches high. The soft cover had a shiny texture, and it bent gently but firmly as I turned the pages. The feel of the book was so concise and contained that its contents seemed all the more foreign: Bashō had wandered through a world not yet touched by the Industrial Revolution.

I could see through the screen two antelope grazing a few hundred feet away. More meadowlarks and buntings darted in song. I turned to the first page of Bashō's story of his journey: "The moon and sun are eternal travelers. Even the years wander on. A lifetime adrift in a boat, or in old age leading a tired horse into the years, every day is a journey, and the journey itself is home. From the earliest times there have always been some who perished along the road."

Travel and time spun, too, outside my tent. What had for millennia been range roamed by people hunting and gathering—signs of human life found in the Powder River Basin are some six thousand years old—gave way just two centuries ago to soldiers staking forts and herders grazing cattle. Then, in 1917, a handful of families came to make a home on this particular piece of the plains. That year, Rhoda and William Mackey traveled south from Gillette in a horse-drawn wagon with plans to turn the prairie into fields of corn, oats, beans, wheat, and more. More than a decade earlier, before they'd set out from back East, the couple had posed for a photo, she with a fine scarf tight around her neck, he with a necktie set in a starched collar. Their eyes searched from stone faces as they squared toward the camera.

For a few years the Mackeys succeeded in making a new life in Wyoming, tinkering with dry-farming techniques to make fertile the small patch of arid terrain. Eventually, five children—Joe, Bess, Rose, Bob, and Elsie—joined them in the wooden homestead with dirt floors they had built just a mile or so east of my tent site. Other families staked claims too, and in the 1920s there was, just north of the field where I camped, a post office, a community hall, and a school. The electrified world the settlers had left behind in eastern cities had not yet arrived. Strong backs and sturdy hands milled grain and hauled and chopped wood to heat homes.

But drought in the Powder River Basin and economic collapse far away proved too much. In the 1930s, not even twenty years after the community, called Teckla, took root, the federal government began buying the land, and people moved away. All told, hundreds of miles of fencing and more than a hundred buildings were torn down. The grasslands were then used mostly for grazing cattle. So when demand for coal climbed in the 1970s, the government was in position to lease land to coal companies hoping to power distant places. As companies came from afar, the earth that held the old Mackey Ranch was dug away. And on the morning that I sat in my tent, there was, where the Mackey kids once planted crops, only the air that filled the depths of the created canyon.

Yet, after decades of digging had opened up the Powder River Basin, coal was falling out of favor in the years before my walk. Power plants across the country burned one-third less coal in 2015 than eight years earlier, with most of the difference made up by natural gas, cheaper and more widely available with the fracking boom. Coal production nationwide had dropped nearly by half in the same eight-year period, and Powder River Basin mines were producing less coal at the time of my walk than they had in any year since the 1990s. Wall Street investors were selling stakes in coal companies, and stocks of several—including Arch and Peabody—were tumbling toward bankruptcy.

A second translation of the verse Bashō posted by his door reads,

> My old grass hut
> lived in now by another generation
> is decked out with dolls

What, then, to make of the industrial world?

In the wandering of ideas across time like those carried by Bashō on his walk, it seemed momentum against coal could eventually silence the big machines. Then, perhaps, another generation could fashion a new, more sustainable system to fuel modern life. As I readied to walk another twelve miles through another hot day, I could imagine an end to the sunken world around me.

When I stepped from the tent at seven o'clock, my pot of tea drunk and several pages of Bashō read, the canyon of the North Antelope Rochelle Mine was opening wider in full light, and the patch of pasture seemed more precariously perched on the abyss. L.J. told me before I started walking that he had lost grazing access to more than six thousand acres turned over to coal mining and that this pasture was the last piece of Forest Service land he was leasing for his cows. I felt peace in the pasture, a temporary resident of a place so lost to time, as though it did not matter whether the pasture was empty or inhabited, whether it existed or not. The canyon of the mine was opening to the south and east. How long would it keep roaming and reaching into untouched terrain?

While I packed up camp, occasional coal-mine traffic traveled County Road 37, as the turnoff to the main entrance to North Antelope Rochelle was just two miles south. The road sat slightly higher than the pasture, so I had to look up at an angle to see the cars and trucks and shuttle buses speeding at fifty or sixty miles per hour, and the drivers, when they turned their heads in my general direction, had to look down. A tractor trailer passed as I was walking from the pasture fence back toward my tent. I was wearing shorts but no shirt, and my belongings were scattered alongside the fence. I had left a pair of orange trail shoes to dry by the road. I must have looked like a homesteader gone mad. Or perhaps the passing driver did not see me at all.

I heard a train approaching. I turned toward the tracks and wandered bare-backed across the pasture. As the diesel engines came even to me, I could see they pulled not coal cars, but flat-beds, and on them something smooth and long and white. I had seen dozens of trains during my days of walking, and none carrying anything other than coal. It took a minute for me to realize: each flatbed was carrying a single blade of a windmill. The parts were traversing coal country en route to some distant wind farm rising in Montana, or North Dakota, or maybe even elsewhere in the Powder River Basin. The blades, each more than fifty feet long, traveled, one after another, with a resolution that was un-expected. I watched for five minutes as the blades rolled by, and then they were gone. The dominant sound in the pasture again came from the steady gusting of wind.

I wondered about Bashō's grass hut and the transformation it might make. One train carrying windmills cannot replace these canyons of coal. But there are many options to replace fossil fuels, from nuclear power and emerging efforts to harness hydrogen to the untapped energy coursing across the earth every day in water, wind, and sunlight. It had taken centuries to build the modern world on the strength of fossil fuels, and it would no doubt be a long journey toward noncarbon sources to replace them.

Was there enough time?

To avoid future catastrophe, experts suggested limiting global warming to no more than 3.6 degrees Fahrenheit. A report from the Intergovernmental Panel on Climate Change a year before my walk had set a harrowing target to stay below that threshold. Between 1870 and 2100, the IPCC projected, humans could add a total of 2,100 gigatons of carbon dioxide into the atmosphere. Studies showed that the current pace would emit 8,200 gigatons, nearly four times that amount. A simulation model prepared by Climate Interactive and MIT's Sloan School of Management offered a sobering assessment: to meet the IPCC target, the world's nations would need to stop burning fossil fuels within the next few decades.

The stakes of the situation mounted with every load of coal that rolled out of the Powder River Basin. The more carbon dioxide emitted by humans in the coming years, the higher the earth's temperature will increase. A projection by the International Energy Agency found that, if current energy trends continue, the earth's temperature will rise more than seven degrees by the end of this century. Such a jump would bring heat waves, drought, downpours and flooding, hurricanes and other extreme weather, to spots around the globe. Those events would threaten the food-and-water system, energy production and supply, transportation and infrastructure, and more. All this would be accelerated, as rising temperatures at the poles would thaw ice and permafrost, which has the potential to release more carbon dioxide than the burning of all fossil fuels remaining on the planet. In other words, the urgency of the switch from fossil fuels, as one key part of decreasing the damage, requires that humans develop renewable, noncarbon sources of energy as quickly as possible.

Yet, standing among the ever-opening coal pits, I could only imagine the wind-energy project toward which the train with the windmill blades was rumbling. I did not know where it would stop or what was happening there. But I did know any such renewable industry could only fully flourish if the mines stopped claiming coal from the ground over which I would walk in the days ahead.

I pushed a few last items into my pack and cinched it shut. A hefty, honking Canada goose began to circle overhead. It was the only goose I'd seen since I set out walking, and it flapped its wings hurriedly to make wide turns, as if it were lost or looking for direction from others. The goose traced a wide arc above me. Its steady honks heaved loneliness into the air. No other goose arrived to honk back, so the big bird circled two, three, and four times. Then it turned north, moving on, and so did I.

※

Three miles up and over a low hill, and the North Antelope Rochelle Mine had already disappeared behind me. But when I turned right at a rare intersection with two lanes of pavement heading straight east, I found myself on another thin edge: the chasm of the Black Thunder Mine loomed along the north side of the road. I knew the Tracy Ranch, the only inhabited outpost I'd heard of in this swath of coal country, lay seven miles ahead. I had called Dan Tracy before setting out. He seemed, in our brief conversation, to be a man of few words and was little interested in me or my walk. But he agreed that I could stop in at the ranch when I passed, and I hoped to be able to camp there for the night.

I checked the Forest Service map again, and it showed the road I was on passing right by the ranch. For one mile, then two, all was well. But then I saw in the distance the Black Thunder canyon had been dug farther south, removing the road I had planned to keep walking on. I would be stuck, without time or energy to detour on foot back around the mine, a trip of some forty miles. I had no option but to continue forward, and I kept a steady pace. I saw a Wyoming license plate that had landed in the ditch by the road. It had the colorful illustration of a rider on a bucking bronco. The plate was marked "TRUCK," and the registration tags had not expired. It was slightly battered, but I hung the plate horizontally on the back of my pack. It made for an odd joke for anyone who passed me. Soon, though, I saw a yellow highway sign in the shape of a diamond: "Road Closed Ahead."

I had decided before setting out that I would move through coal country without needing anything from the mines or miners themselves. Company officials had denied my requests to visit both North Antelope Rochelle and Black Thunder Mines after weeks of negotiation by email, and that was okay. It felt better to move among the mines as an outsider, aware of the greater toll of the harvest, rather than as someone trying to make sense of the system from within. But, with the road ahead of me removed by the mine, I needed at least a small kindness from those on the inside.

I came to another mine entrance that angled off the road, and there stood a small guard shack. I walked up to the building and stopped where a vehicle would stop, in front of a closed glass window. I looked through and saw no one. I knocked. No response. I walked to the back, looking for an entrance, and found a woman smoking a cigarette. She turned to me and said, "You come inside and cool off." As we stepped into the small building, she pointed to a room on the right that had lockers and a utility sink. "You can grab that chair and pull it in here by mine," she told me. I slid the seat into the bigger room, where she sat at a desk alongside the window I'd knocked on. She handed me a Styrofoam cup and pointed at a cooler. "Help yourself to water."

We quickly set to telling each other stories, she about her years shuttling conductors out to the coal trains, me about my walk up to her world. Her name was Sindy, spelled, she pointed out, with an S, and she told me about her granddaughter in Douglas and the two-hour drive she makes into South Dakota to visit her mom, who lived alone. A big rig pulled up at the window, and Sindy slid it open and threw out a warm "How you doing?"

The driver told Sindy the name of his company. "I need to mark this down," she said, turning to her ledger. There was another window behind me, for vehicles leaving the mine, and a pickup truck pulled up there. Two miners laughed as they told Sindy they had been washing dump trucks all morning. Their faces were caked in dirt.

The trucks drove off, and Sindy pulled out a tablet computer and began to play *Candy Crush Saga*, the little towers of sweets spinning across the screen. I plugged in my iPhone and charged a backup battery I carried. I dug in my pack and ate one granola bar, then some jerky, and got hungry for more. I lost my gumption to move.

"You stay as long as you'd like," Sindy said.

It was only a few hours since I'd stood beneath the circling goose at my pasture camp, but I reverted so quickly to the comforts of the electrified life. The guard shack, too, seemed suddenly like home. As more trucks pulled up to Sindy's window, more candies crushed across the screen, more minutes passed without feeling the heat of the pounding sun, I became fully an insider again, insulated from the natural world. I was so conditioned to live in predictable comfort, and it felt good not to have to work so hard in the heat.

After two hours of rest I knew that if I did not move it would only get harder. I asked Sindy if somehow I could cut through the mine to the Tracy Ranch. She handed me a map and pointed out the window toward a dirt road that led into a wide pit. "No problem," she said. She set down her tablet and pivoted in her chair. "Time for a cigarette."

The trucks coming and going from Sindy's guard shack were heading south or west, and I east. I stepped back into sunlight, my shoes scuffing across the gravel lot, and just like that I was alone again, wandering the red-dirt road that Sindy had suggested, passing through a canyon of the mine that was otherwise empty, as work occurred elsewhere. The road descended gently as the rim rose above my head, and I found myself surrounded by the vacuum created by the quest for coal. All was rugged rock and dry dirt, and I stopped, overwhelmed by the emptiness. The road itself was freshly graded, benches of earth still holding the shape given them by dozers and graders and trucks. It was hard to tell exactly where the coal seam had run, but I wondered how much energy that source had provided, how many days of distant comfort had come from this single transaction. Already the big machines had moved south, deeper into another canyon, in search of more.

I climbed the far side of the canyon and then picked up the public road on the other side. Around a rare notch of low hills, I came to a dirt driveway that ran off at an angle toward green-roofed buildings and a lush lawn beneath strong trees: the Tracy Ranch. Three generations lived there. Jake, the grandson in his twenties, was off in an oil field checking on wells. Jacci, Jake's mom, somewhere around fifty, had just been puttering on an ATV, working calves. And Joyce, Jacci's mom, was taking a seat on the patio in a sturdy iron rocking chair that Jake had made by welding together dozens of old horseshoes.

"You're going to have horseshoe butt," Jacci joked as Joyce sat down.

It was just after four o'clock, and I plopped in the shade next to Joyce. Dan, her husband, was due back from the dentist in Newcastle. He'd broken some teeth days before while out on an ATV, chasing cows.

Joyce set to rocking. Her mind seemed to drift, and then she perked up. "I just love it here," she said. She told me she had known that when she first set eyes on the place twenty years ago. She and Dan had been looking for a ranch of their own, so they moved up from Colorado. The mines were distant neighbors then.

I was tempted to talk about the ever-expanding canyon, which now reached to within a mile, just up and over a hill. Of course, Joyce could hear the grind of the big trucks through day and night and the sirens warning of a blast. She could watch as a cloud rose after each alarm, sometimes floating over the ranch.

But the Tracys' increasing isolation alongside the expanding mine seemed so inevitable that I just sat and listened as Joyce went back even further in time, telling of the family that had homesteaded this ranch a century before. A woman in that family had planted cottonwoods, and each day she hauled water from School Creek, which flowed a few hundred feet west of where we were sitting. "She carried seven buckets a day," Joyce said, announcing a precision of fact or fiction that had passed through generations. "She must have loved those trees." More than a dozen cottonwoods, their trunks thick and tall, branches arcing in an umbrella of shade, towered above us. Joyce looked at them, her eyes seeming to measure both memory and soft light at end of day. "It is such a beautiful place," she said.

During the next hour, a storm blew in, dumping sheets of sideways rain as Dan arrived home. Then wind carried the clouds away, and by midnight silent stars sat on the horizon. Joyce and Dan were tucked in their small ranch house at one end of the yard, Jacci and Jake in a house at the other. I lay in my tent on the lawn between. Strong gusts came constantly. Tex, the cattle dog, charged past time and again, barking at something that alarmed him. The cottonwoods creaked overhead, their limbs reaching down in the dark.

※

I knew even before I made the turn from School Creek Road the final miles would be particularly intense. I would camp one more night, at a reservoir on Little Thunder Creek, but to get there I had to walk ten miles from the Tracy Ranch through the center of the Black Thunder Mine. State Highway 450 ran between northern and southern canyons created to claim coal. Earlier in the walk there had been enough stillness for me to observe from the edge, as though uninvolved in what was happening in the mines below. But along 450 I was exposed and at risk. There was coal-mine traffic, but also tractor trailers hauling out to oil fields, locals heading between Wright and Newcastle, and occasional tourists crossing the basin between the Black Hills and Big Horn Mountains. Unexpected thoughts arrive when moving close to the ground, all soft flesh and carrying nothing but a walking stick. What if a fire extinguisher strapped to the back of a pickup truck bounced off and hit me in the head?

Despite fatigue from earlier days of walking, I kept a brisk pace and didn't break stride for more than an hour. Then, panting dry breath in the rising heat, I stepped a few feet down a slope along the side of the highway, dropped my pack, and sat. I was looking south, across pasture that ended at an open pit just opposite the east entrance to the Black Thunder Mine. The view held a confounding mix of order and chaos, a field of brittle grass giving way to the cluttered coal pit. After days of moving amid such altered earth the scene felt familiar, as though there were no other way it could be.

A white pickup truck pulled over behind me, and a young woman wearing a hard hat and orange safety vest called out. She reached a tattooed arm out the window, offering a cold can of Pepsi. I had left the Tracy Ranch before Dan and Joyce had emerged from the house, so the talking I had done that morning was to myself. Cumulative days alone messed with my sense of social engagement. I walked up to the truck and took the Pepsi from the woman's outstretched hand.

"Perfect timing," I said, showing her an empty granola bar wrapper. "I needed to wash this down. Thanks."

That was all I said. Nine words. Then I turned back toward my pack and sat, the woman and three others watching me as though

I were an animal in a wildlife preserve. I slugged back the Pepsi in two or three gulps. The infusion of carbonated corn syrup jolted me. The truck pulled away and turned into the mine, and I walked west. I felt confidence in my independence, as though moving slowly, step-by-step, through a place long seen as only a resource gave me authority about what might happen there.

I was not so naive to think shuttering the mines would be quick or complete anytime soon. The presidential election the year after I passed through the Powder River Basin would reverse course on some of coal's decline. The Trump administration would withdraw a Department of the Interior order that had placed a moratorium on future coal leases on federal land, and it embraced a general policy of coal expansion, envisioning fossil fuels as the key to the country's energy future. But the broader collapse of coal could not be stopped by politics. An investigation by the Reuters news agency found that, since 2010, more than half of the 523 coal-fired power plants in the country had gone out of operation. Utilities were investing in natural-gas plants, and wind- and solar-energy systems, particularly, were becoming cheaper and easier to build. It was incremental change, to be sure. But the vast energy system was shifting away from the most carbon-intensive of its fuel. Dirty and tired, wind-whipped and hot, I knew that if the public and its policy makers could choose to invest in renewable sources—water, wind, sun, or something else—then the Powder River Basin could return to its original purpose.

I shouldered my pack and tried again to pick up my pace. Just past the Black Thunder Mine's east entrance, I heard a beep from behind, and the same pickup truck that had pulled over sped past. The driver gave a big wave. I held my walking stick above my head and pumped it up and down in confident greeting, stronger from the certainty of my emotions.

Four miles later, at the western end of canyons already created, I stopped on a bridge where Highway 450 crosses more train tracks. Just north, a single train track passes through a tower that has for years loaded more coal than any other in the Powder River Basin. I sat with my back against a guardrail post and settled for six inches of shade from the midday sun. The loading tower was quiet, and it looked to me to be an abandoned relic of a long-dead industry. I could imagine the steel of the tower dropped back to earth, rotting into the soil like a fallen tree into a swamp so many millions of years ago.

⁂

Traffic on highway 450 kept hurtling along, and soon I was walking the final stretch to Little Thunder Creek and a reservoir made there nearly a century before. I turned south on another paved road. Heat and fatigue conspired to create a rash between my legs. I hobbled onto a dirt track that led west into sagebrush and bent grass. All the Black Thunder Mine was behind me.

The only tree for miles around stood at the far shore of the reservoir. I descended into the dry creek bed beneath a dam built soon after homesteaders had begun to flee Teckla. I climbed the other side, and a few hundred yards from the tree I saw, lying between two sage bushes, the first dead rabbit. The rabbit was more than a foot long and had soft white fur. I did not pick it up or move it at all, but I could see no signs that the rabbit had been attacked or visited by scavengers. How could such food not be consumed in so barren a place? Twenty feet later, off to the left, I saw the second dead rabbit, and then a third.

I arrived at the reservoir's edge and saw that the tree was little more than a stout shrub, with tight silver-green leaves that flipped in the breeze. I sat close to the narrow trunk for a few minutes of shade. Dry sweat caked my neck, and my shirt hung wet against my sides. Two live rabbits, one gray, the other white with black and brown spots, ate thick grass against the shore. Murky water rippled. The rabbits stared at me, noses twitching, and I at them.

They must have been descendants of domestic rabbits dumped here. But when, and by whom?

It was midafternoon, and the high heat was relentless. The shade from the tree's branches shifted out over the water as the day drew on, so I set up my tent and crawled in for relief. I stripped down and lay on my sleeping mat. I stared at the blue sky. I could hear a coal train rolling south. Coal may decline, but what would replace it? The fracking industry had only expanded the use of natural gas in its place. With easy energy cheaply available from other fossil fuels, could renewable sources, such as water, wind, and sun, ever grow at a rate to provide as much energy? I thought of that single train that had passed my campsite the morning before, its flatbed cars carrying the giant white blades for windmills, and I wondered again where it was going.

That last afternoon alone near the coal mines marked a threshold between place and time. L.J. had agreed to drive thirty miles from his ranch to pick me up at the reservoir the next morning. I was ready to return to New Hampshire and my family, my mind and body exhausted from this reckoning of fossil-fuel landscapes —oil in North Dakota, gas in New York and Pennsylvania, and coal in Wyoming. I had seen so clearly the power of the existing system and the inertia of daily need for fossil fuel trumping any ideals about combating climate change. But I also knew there had been much happening to develop renewable fuel at an industrial scale during recent decades, and I was optimistic about my next steps. After returning home to New Hampshire, I hoped to walk on to a second stage of my journey, moving toward the future: along the powerful tides of coastal Maine, in the wind-whipped plains of Texas, and across the sun-scorched Mojave Desert of California. I was curious to know what renewable landscapes of fuel might feel like, and I was eager to walk toward the potential of water, wind, and sun.

I woke from a doze in late afternoon to see a smaller white

rabbit, another member of the family living and dying amid the sagebrush, sitting a foot from my tent door. Its nose twitched. I was sweating in the still heat. The rabbit and I were not suited to this compromised terrain. I told it to move along.

Spring storms had drenched the prairie for days before I set out walking. I decided to check my phone's weather app to see if there would be rain the next morning. When the app opened, I saw a thick band of red, and in that band the words "SEVERE THUNDERSTORM WARNING." I touched the words and more details emerged: trained weather observers had reported winds of sixty miles per hour and quarter-sized hail near the town of Spotted Horse. I opened my map and found Spotted Horse northwest of Gillette. I read the warning again. The storm was moving fifty miles per hour to the southeast, directly toward me. The warning advised anyone in the storm's path to seek shelter in an interior room on the ground floor of a building.

I climbed out of my tent. Blue sky still opened east. I turned around, and to the north and west all was black as night. Several miles away a white curtain of hail hung just above the ground. Lightning split the scene. I looked down at my tent. A gentle breeze still arrived from the east. I climbed inside and zipped shut the fly. When the wind shifted, would the storm throw the tent, and me, into the water? I opened a knife and kept it in reach, in case I had to cut my way out. More likely: I'd get pummeled in place. I lay face down and set my still-rolled sleeping bag over my head. I said out loud: "I love you, Julie. I love you, Luca. I love you, Colette."

Within minutes strong gusts swept in from the northwest. Thunder growled, then slammed. The hail arrived first in volleys, then with solid force. The tent walls snapped beneath blasts of hail, lashing against my back. Water rained in at the edges. For one minute, then two, I lay stock still, counting my breath against the storm. I remained face down, observing the scene through sound. The low moan of a coal train rose from the east, and there came all around this audible collision: growl of coal cars and howl of whipping wind sparring above the prairie. Eternity compressed into moments, as the train and thunder rumbled. I felt unexpected comfort in the face of nature's force, as though my exposure ensured I would survive.

Slowly, as the train moved south and the storm east, the clash quieted. Last bursts of hail spattered, and the air calmed. I rolled onto my back and unzipped the tent door. It flopped in light breeze. The temperature had dropped ten degrees or more. Sagebrush branches braced against cuffs of ice. A thick scatter of hail lay on the ground. I climbed out of the tent and took my first steps forward. The still-trembling earth crunched beneath my feet.

Part 2

# On to Tomorrow

# Chapter 4
# Idle Currents

**I stepped, unsure, from one slick rock to the next,** picking my way toward the center of the St. Croix River. There, at the eastern edge of Maine, I balanced atop a piece of granite about the size of a beach ball, the water only inches away. A steady rush of air rose from the river's surface, as clear currents swept between stones and spun in small eddies before running again toward the Atlantic Ocean.

This single channel of water, so consuming up close, swept through the forested bend as an imponderably small sample of the water on and within the earth. Only a fraction of 1 percent of all water moves on the surface in lakes, ponds, swamps, rivers, and streams. Most of that travels back to the ocean, home to nearly 96 percent of water on the planet. Surrounded by the St. Croix, I was absorbed by the cycle never-ending: gravity pulling the water into the ground or toward the sea, where it would evaporate upward, the sun's heat hoisting it higher, only to condense into clouds and fall again, a natural momentum, from one rock to the next, one year to the next.

After my three walks across landscapes of fossil fuel, I was drawn to the watery edge of the continent to reconnect to natural currents and explore what might power the industrial world into the future. During those November days of 2015, I planned to follow the St. Croix's course twenty-seven miles downstream to the city of Eastport, nestled against the Bay of Fundy, home to some of the highest tides in the world. There a band of tinkerers had been experimenting with turbines that, submerged, could capture the power of the rising and falling sea to generate electricity onshore. Tides hold massive potential to power a significant portion of electricity. Innovative projects—using floating barrages to hem in the sea, or sunken turbines—have been built in recent decades. But they receive relatively little funding and investment, so the tides are so far largely untapped. The Eastport project, which ran from 2012 to 2014, had been the first in North America to deliver tidal power into the electric grid. My journey toward the sea, then, would offer a chance to consider one small attempt to realign the industrial system on renewable, noncarbon fuels.

The sun, the original source for all the carbon buried deep underground during millions of years as oil, gas, and coal, creates currents of existing energy nonstop, whether with constant rays, wind lifted by warming air, or passage of water through the hydrologic cycle. In the decade before my arrival in Maine, solar and wind power had been growing rapidly as a source of electricity, and I hoped to walk later through the wind farms of Texas and the largest solar plant in the United States, in the Mojave Desert of California. First, though, I had come to the rocky coast to consider the potential of the tides.

The Milltown Dam, less than a mile upstream from where I

stood in the St. Croix, was one of the first in North America to use the power of water moving downstream in a river to produce electricity. On Christmas Eve in 1887, a rope-drive turbine-generator powered by the St. Croix's currents sent electricity to Cotton Mill Lane, on the Canadian side of the river, where the energy lit a lamp. Six years later, at the corner of Queen and Church Streets, the home of mill manager Lewis Dexter became the first to receive electricity. By 1920 a new powerhouse with three turbines and generators was sending electricity from the mill compound—by then a network of concrete dams and towering brick buildings set above huge granite rocks—to surrounding neighborhoods. During those and following decades, dozens of other dams were built in the St. Croix River basin. That was part of an expansion of hundreds of hydroelectric plants across the United States that included, in the west, the massive Hoover, Grand Coulee, and Chief Joseph Dams. Modern turbines at the Milltown Dam still deliver electricity into the grid, feeding both sides of the border. And today river-based hydropower fuels 6.5 percent of all electricity consumed in the country. But river dams damage ecosystems, and after decades of debate there is little interest in or effort toward building more.

Downstream from Calais, around Eastport, the small group of engineers with the Ocean Renewable Power Company had tried to develop something less intrusive: turbines that, suspended underwater from the sea floor, harness the power of water rushing in and out with the twice daily tides. It was not a completely successful attempt, but it provided valuable lessons the team continues to build on in other tidal generation projects. That seemed to me exactly the spirit needed for the broader challenge of reengineering the industrial grid on renewable fuels.

On that first frigid November afternoon, it felt good to be surrounded by the St. Croix, exposed again in nature, and I hoped in this second stage of my walks to gain insight into a cultural question: can humans, as a collective society, act again as animals dependent on natural rhythms? I felt in the cadence of the St. Croix's current a perpetual certainty. I stooped slowly and dipped one hand into the clear shiver of the soon-to-be sea.

❋

The next day, three miles downstream, I slumped my backpack from my shoulders in a roadside park and sat on a bench to look at the low-tide wash of the St. Croix. The river had widened to 1,500 feet across, but in early afternoon the flow was reduced to a much narrower channel, flanked by mudflats. The day was quiet gray, with a slight wind blowing downriver, cooling even more the forty-degree air.

My pack was fifteen pounds lighter than on earlier trips. I carried no tent, nor sleeping bag, as I would spend nights either in motels or private homes as I passed from Calais to Eastport. I had no cook stove, and hauled only a half gallon of water, sure that I could stop at a house for more. I did have two rain jackets and extra changes of underwear, in case I got soaked through.

The stretch of U.S. Route 1 I'd been following east along the river had an almost suburban feel, passing mowed lawns and swaths of small forest that had long since been cut of their original timber. But my perch in the park looked out over a world still wild: the rugged river and its banks, where the twice-daily tide returned everything to its original state. A young bald eagle, its head still brown, soared above the mudflats, drifting in the wind on tilted wings. A stiff gust swept the bird above evergreens, where it settled.

For millennia native peoples living near the St. Croix had shifted their lives with the seasons, paddling between the Atlantic Ocean and inland waterways, as the earth's orbit around the sun turned summer to winter and back again. Groups of Passamaquoddy, as the local tribe is called, settled along rivers and the coastline during spring and summer months, spearing salmon, digging clams, and hunting seals with little concern for warmth or stores of provisions. Any sustenance and survival during the easy season came one day to the next. But with autumn, and the turn toward darker days and colder temperatures, the tribe would move inland, following the St. Croix to a sandy beach on the bank opposite modern-day Calais.

On my first day in Maine, I'd met Donald Sactomah, a Passamaquoddy historian who still paddles the St. Croix's strong currents. We'd stood in the center of Calais, and Donald had pointed across the river to the site of the sandy beach, saying aloud a few words in Algonquin, the traditional language of the Passamaquoddy and other native people in the region. Then he offered a translation. "The words," Donald told me, "mean 'The Great Landing Place.'" From that sandy landing they would paddle and portage many miles upstream through dense forest to make winter camp alongside frozen ponds and lakes. Sturdy wigwams, each with thirteen birch poles lashed together with spruce roots, offered shelter. The thirteen poles matched the thirteen full moons that appear each year, bringing the tides even higher. Each wigwam was shaped in a circle, with the door facing east, in the direction of the rising sun, the source of survival. Within a wigwam's walls, as a wood fire burned for cooking and heat, smoke rose toward a hole at the top, ash drifting skyward. When the returning sun melted ice each spring, the Passamaquoddy moved again toward the sea.

I picked up my backpack and stepped steadily along the shoulder of Route 1 as it arced up a rise. Cars and trucks hurtled along the two-lane road at sixty and seventy miles per hour, each able to cover the twenty-seven miles of my walk to Eastport in less than half an hour. Most vehicles had only one person inside, an inefficient arrangement for moving so much mass between two points. I knew from my own experience driving at home the sense of hurry maneuvering a machine a mile a minute brings. Driving,

I am often consumed more with getting to the destination than with being focused on where I am. Walking along the roadside, I was happy to be moving again in the immediate moment. One foot in front of the other, each step carried me within the limits that it could.

Six miles and two hours east of Calais, I came to a small parking lot at the edge of a forested peninsula called Devil's Head, which juts into the St. Croix where the river widens to a mile across on its journey to Passamaquoddy Bay. I walked among the trees and hid my pack behind a trunk, then strolled unburdened along a rutted road until a track descended toward the riverbank. More than a century before, long after Eastport had been established as a deep-water harbor, a seasonal hotel had been built on Devil's Head. Steamers moving between Eastport and Calais stopped, and travelers disembarked for tea and a snack.

The hotel is gone, but I sat on a rock near the old dock site and drank deeply from my water bottle, looking across the dark surface of the river at brown fields on the opposite slopes in Canada. The expanse was as vast as that I'd navigated on the prairie of North Dakota and Wyoming. I fixed my gaze three miles downstream, at a small swath of land in the middle of the river: Saint Croix Island. The island measures only six acres, with sand and rocks edging a low rise of soil capped by pine, and it tilts at an angle in the river. The tide was coming in, and the wind stacked whitecaps, so the island seemed a stout anchor amid the wash. I could see in its position both safety and isolation, as water rocked all around.

In June 1604, fifteen years before the English arrived in Jamestown, a company of French explorers sailed up the St. Croix and chose to make a settlement on the small island. The explorers were traveling on orders of the king to spread Christianity and claim the region. In his journals Samuel Champlain, mapmaker on the voyage, explained the choice of Saint Croix Island as easy:

"the best we had seen both on account of its situation, the fine country, and the trade we were expecting with the Indians of these coasts and the interior, since we should be in their midst."

During the summer months the crew built a fort, a storehouse, and cabins. They pulled alewives and other fish from the current and pried shellfish from mud at low tide. The Passamaquoddy, come for fishing during that season, moved easily among them. The French, noting that the region was about the same latitude as their homeland, counted on a mild winter. They did not expect the Arctic air arriving from the north. The Passamaquoddy paddled back upriver as autumn arrived. Soon the French found themselves stranded on the island: the relentless tide kicked up solid sheets of ice, rendering the river too dangerous to cross.

By the time the Passamaquoddy had broken camp the following spring and moved from their wigwams in the interior woodlands toward the coast, more than thirty-five in the group of seventy-nine French explorers had died on Saint Croix Island. "And more than twenty were very near it," Champlain wrote. Ships arrived from France in June, and the survivors moved on, with Pierre Dugua, the leader, making a new settlement eighty miles east, in present-day Nova Scotia. Before leaving Saint Croix Island, though, the French dismantled buildings and hauled their parts and anything else of value to the new settlement, called Port Royal. Even in those early months of 1605, their departure from the St. Croix River must have seemed only temporary, as the explorers had measured the bounty of both land and sea, endless resources for a world accelerating toward the Industrial Revolution.

I rested on the rock on a dank day four centuries later, that revolution having long since turned to industrial rule. I watched the rising tide cover several one-foot-high stones in the span of thirty minutes, and distant Saint Croix Island, its shores shrinking with the deepening sea, shuddered as a moment in time: 1604, that fulcrum between one world that long had been and another arriving

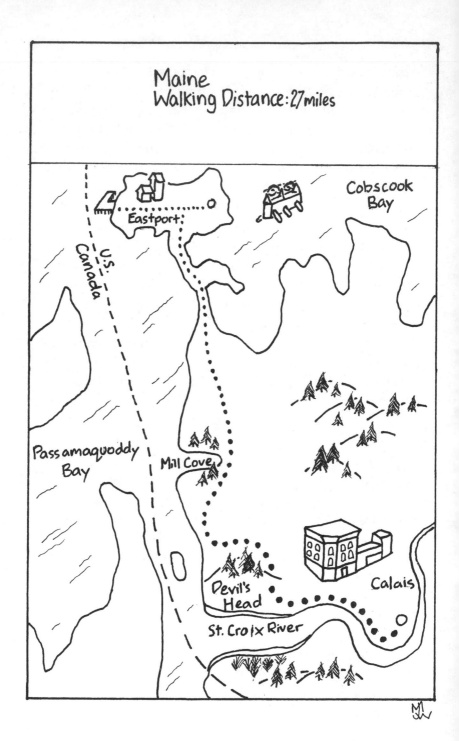

hungry for more. English settlers had soon followed the French, and centuries of local economy along the St. Croix—of fabric, fish, and forest—ran their course as communities harvested in a hurry, only to be outpaced by global markets in recent decades.

By the time of my walk, the coastal countryside of Maine wallowed in a sort of slack tide, the biggest businesses a hospital in Calais, a border-patrol outpost, and the service stations selling cheaper gasoline to Canadians coming south. The sardine industry that had flourished along the St. Croix and around Passamaquoddy Bay a century earlier had long since packed up, and the town of Eastport, downstream, staked its claim more on tourism than industry. Saint Croix Island, maintained as an International Historic Site by the National Park Service, is off-limits to visitors, and when I walked past the visitors' center on the mainland, an hour after moving on from Devil's Head, the parking lot was empty and a small office building closed, the season still sending the message to move on.

I continued step-by-step along Route 1, and I was struck more than on any other walk by the accumulated litter, settled like fallen leaves along the shoulder of the road. Here is a sampling from an incomplete list: a bent can of diet A&W Root Beer; a single CD from volume 2 of *Big Band Hits: Gold Edition*; an empty pack of Marlboro Gold cigarettes; a pack of Camels; so many cardboard cups, the kind sold in gas stations to hold French fries. The refuse continued, mile after mile, sometimes as a single piece, other times in vast clumps, as flotsam bobbing on a still wave: a Mr. Goodbar wrapper; a Circle K Froster Styrofoam cup; a blue rubber dish glove; an Oh Henry! bar wrapper; the handle of a hatchet; a piece of rope; several fashion magazines, with the covers torn off; a McCafé cup; a Bud Light can; a piece of plywood; a cup holder with the Green Mountain Coffee Roasters logo; a wrapper from a Little Debbie's Peanut Butter Creme Pie; the other blue rubber glove (this with the Atlas Vinyl Glove logo still visible). Most common, with more than a dozen along a stretch of a few miles, were long plastic sleeves that once held Slim Jim beef jerky. Was just one Slim Jim fanatic flinging a wrapper a day while speeding past? Even for me, moving on foot, the litter began to blend into the environment. Passing drivers likely could not see it at all.

At my slow pace, covering three miles per hour beneath the pack, the roadside trash rose up as another measure of how modern consumption has grown out of proportion to the planet. Each day people living in the industrial world generate millions of tons of trash. And I wondered, walking alone as the cars whistled past on Route 1, if enough carbon-free fuel could ever be harvested to provide all that modern life demands, let alone tosses away.

Halfway on my walk between Calais and Eastport, the St. Croix widened into Passamaquoddy Bay. The bay stretched east and north for ten miles from Mill Cove, a small notch in the riverbank that had been carved by a stream flowing from inland hills. When I arrived late in the afternoon, the cove was full with the high tide that had risen during my walk. Soft light shone on the calm surface of the bay, and everywhere were gentle hues of blue.

❁

On coastlines all around the world, twice a day, the gravitational pull of the moon and the countering power of the sun shape the rise and fall of ocean tides. The range of those tides—with high and low six hours and twelve minutes apart—are determined by the geography of the sea floor and the force with which water moves across it. Along narrow coasts those factors are amplified, and the water rises particularly high.

The drop between high and low tide in most coastlines is not enough to generate significant power to drive turbines and produce electricity. But where that drop is higher than eighteen feet, nature's surge and recoil carries with it displacement of the water that can produce power. Such spots exist along the northern reaches of Russia and the east coast of Europe, in Australia and south Asia. In North America nowhere does the water rise and fall with such depth as in the long, narrow Bay of Fundy and just to the west, tucked between the mainland and a large island, in Passamaquoddy Bay. By nine o'clock the next morning, roughly eighteen hours after I had arrived at Mill Cove, the previous afternoon's high tide had fallen and risen and fallen again. So I saw not the shimmering surface of high water but mudflats laced with trickling streams for hundreds of yards offshore.

This impressive twice-daily migration of water had caught the imagination of Dexter P. Cooper a century earlier, when the engineer hatched a plan for harnessing the bay's power. Producing electric energy would require sending the water from a higher pool into a lower pool, and Cooper envisioned filling the crooked coast along the southern edge of Passamaquoddy Bay with a series of low concrete dams that would seal the water in at high tide. He planned a second series of dams at the eastern edge of the smaller, neighboring Cobscook Bay, and those would be opened to empty the lower pool as the tide rushed out. With the natural passage of the tides interrupted, gates of a power station, set between the high pool and low, could then be opened. The water rushing from Passamaquoddy's high pool to Cobscook Bay's low pool would turn turbines, creating energy that could be shipped along the shore through power lines.

Such a grand scheme required not only capital to build the dams but unprecedented permission from government agencies to reshape the natural order at the edge of Downeast Maine. Cooper's plan got a big boost in the early 1930s, when President Franklin Roosevelt, who knew the tides well from summers at his family cottage on Campobello Island, supported the project. One year and $7 million later, a crew of hundreds of workers had built two long dams along the south end of Passamaquoddy Bay. One ran between the tip of Pleasant Point on the mainland and Carlow Island, and another from Carlow down to Moose Island, home to Eastport. At the north end of Moose Island, a pop-up community had arrived too, with a school, library, dormitories, and single-family homes built for a thousand workers.

The completion of the Passamaquoddy Tidal Power Project, as it was called, was expected to take many more years. But work ground to halt before the end of the decade, as opposition from established utilities in Maine and members of Congress from southern states, who saw it as a Yankee indulgence, refused to fund it further. World War II soon took attention from big endeavors, and crucial dams that would hem in the tides between Moose and Deer Island, and from Deer Island north to New Brunswick, were never built.

Like the pull of the moon, the lure of the Quoddy Dam did not stop, and in July 1963 the Department of the Interior published a report about the feasibility of completing the project—examining costs, power-generation potential, and more—at the request of President John Kennedy. "The time is at hand for America to fulfill man's centuries-old dream of harnessing the energy of the tides," Interior Secretary Stewart Udall wrote at the opening of the report. "The place is Passamaquoddy Bay."

The report projected that after $1 billion of construction the upper and lower pools of Passamaquoddy and Cobscook Bays could generate one million kilowatts of electricity each day during peak demand, feeding the grid and serving customers throughout northern Maine and south toward Boston. The report noted that production costs per kilowatt could be competitive with power plants burning coal. It made the case, too, that local employment and the Quoddy Dam's draw as a tourist attraction should be counted among the project's merits. There was no mention of the need to replace fossil fuels, as the toll of their consumption had not yet been so thoroughly measured. Still, Udall argued that the

United States should make the investment in a system that may not be as easy as burning more coal. "Harnessing the energy of the tides is a big idea and a big undertaking," he wrote, in closing. "We must think big and act big if we are to take full advantage of the opportunities modern technology holds out to us."

Four months after the report was completed, Kennedy was assassinated, the Vietnam War accelerated, and the project stalled again. Interest in the Passamaquoddy project returned briefly a decade later, during the oil crisis of the early 1970s, and again as recently as the 1990s, when Senator Olympia Snowe of Maine requested the Congressional Research Service provide data on the energy potential of the tides and the history of the Passamaquoddy project.

Still, nothing ever came of Cooper's grand design, and by the time I walked south from Mill Cove, Passamaquoddy Bay rose and fell with the pull of the moon. By late afternoon I'd covered five miles along the bay's southern shore, where Pleasant Point extends toward Moose Island. The point is home to a reservation of the Passamaquoddy Tribe, which also has a reservation thirty miles inland, at Princeton, set among lakes. The two locations mirror the tribe's historical seasonal relationship with the sea, though in modern times, with tribal members living without the natural migration of the seasons, the reservations have become outposts of poverty.

I'd been told before walking south that I may not be welcome cutting through tribal land at Pleasant Point, but the first person I met, an elderly man out working in his yard, smiled widely and called out a greeting. He was lean, wearing a Nike baseball cap and an old army jacket, and he turned from raking soil from a ditch to smooth out his already tidy lawn. "Oh, I'd love to do what you're doing right now," he told me as I approached, walking stick in hand. "I've got a canoe, and I've always wanted to go, just me and my dog, and paddle."

"Where to?" I asked.

"Everywhere," he said.

His name was Frank Bassett, and he told me it was his birthday.

"How old?" I asked.

"Guess," he said.

I looked at his nimble frame, the sparkle in his eyes. "Eighty-one," I said.

"Eighty-three," he said. "Born November 18, 1932."

Frank's lawn had an intimate view down a tidal creek that ran toward the end of Pleasant Point. "You're on the reservation now," he said. He had done a lot of wandering himself, from six years in the U.S. Army, to forty-four years at a chemical company in Massachusetts. He'd hiked Zion National Park and trails all through the Rockies. "A lot of people move along so fast they don't know where they are," he told me. "I try to look around." Frank knew he was slowing down, so he had come home to Pleasant Point. He had gout in his legs, and that autumn he had finally given away three pair of cross-country skis. He planned to spend the winter at rest, cozy in his small house, heated through the deepest cold. "When I was a kid, we didn't even have electricity," Frank told me. "Just kerosene lamps and an outhouse. So we've come pretty far, I guess."

It was only three thirty, but already an hour before sundown, and I still had several miles to cover to get to Eastport. Frank invited me to stop back if I ever passed through again, and I picked up my pace heading south, first along a stretch of bike path set close to the water's edge. The sharp slant of sunlight painted everything in soft orange. I cut down a side street past the Pleasant Point school and onto the first causeway built on the 1930s dam that had been a step toward damming Passamaquoddy Bay. Cars sped south toward Eastport, or north toward the mainland, as I traversed Carlow Island and then the second dam-turned-causeway to Moose Island. Darkness was coming quickly, with a stiff chill as the temperature dropped into the thirties.

I passed the old Quoddy housing site, its dormitories and community buildings still standing, but many empty. At a wide bend in the road, Raye's Mustard, once an anchor of the sardine-canning industry and now a specialty shop that touts its status as the only dry-ground mustard manufacturer in the United States, was shuttered for the night. I followed a narrow residential street due east past nestled wooden houses, many built during Eastport's early days as a fishing hub in the 1800s. Eastport had peaked around 1900, when more than a dozen sardine canneries supported a population of 5,300 people. Now summer tourism offered as much promise as any industry, and only 1,300 souls stayed through winter. The city had put some hope in a deep-water harbor, an attempt to capitalize on the coast that had long defined it, but just one or two ships called each month to collect cargo.

It was dark when I stopped outside a house a few hundred yards from Western Passage, which connects Passamaquoddy Bay to Cobscook Bay. Though I could not see the water, I smelled salt in the air, and I knew the outgoing tide was marking time. The elderly woman who ran a bed and breakfast in the house had assured me there would be a room. I knocked on the front door and waited, the only light the glow of a television behind the window.

※

Early in this century the unexpected outfit from Ocean Renewable Power Company arrived in Eastport with their idea for harnessing the twice-daily tides that rise and fall around town. This was not a grand government project to redefine the coastal terrain but a focused effort by the small team of engineers based in Portland. Their goal was the same as Dexter Cooper: deliver kilowatts into the electric grid to power life onshore. But the design was different. Where the Quoddy power project had required containing the sea, ORPC planned to submerge horizontal turbines to capture energy rushing in and out with the tide. Much like a wind turbine rising above ground or a solar array set beneath the sun, the goal was to use the submerged system to capture existing, carbon-free power as it moved across the earth. The challenge was finding materials and construction techniques that could help the turbines and other parts of the system withstand the push and pull of the saltwater bay. A half century after the United States had invested billions to put humans on the moon, the efforts to answer such key questions about harnessing the power of the moon's pull were strikingly limited, with ORPC an innovator at the edge.

The oil and gas industries receive roughly $4 billion in subsidies each year in the form of tax breaks from the cost of their operations. Yet relatively little government money or private investment was being dedicated to tidal innovation. The ORPC team, just three people when they first brought their plans to Eastport in 2007, gained funding through a series of grants, including $300,000 from the Maine Technology Institute and $240,000 from the Massachusetts Technology Collaborative. On the strength of early designs, the company landed $5.5 million in private investment, and then, after several other small grants, the U.S. Department of Energy awarded the company $1 million to design and test a mooring system as part of its project.

Moose Island is roughly four miles long and two miles wide, with the tighter city streets of Eastport set on the southeastern end. Farmland and light industrial outposts sprawl across the western half. The island is shaped like a particularly ambitious piece of a jigsaw puzzle, just a few hundred yards across in the center, with its edges carved in stout peninsulas and sweeping arcs. The west coast of the island is defined by three of the most prominent: Broad, Deep, and Carrying Place Coves. Cobscook Bay sits off the western shore, and the engineers had chosen a deep spot just past Cooper Island Ledge to set their unique cross-flow turbines, designed to withstand the pressures of shallow tidal rivers and seas. At the outset the devices were a work in progress, with different versions made from different materials, cut to different lengths. But they each shared similar characteristics, with sculpted horizontal blades wrapped on an iron frame, not unlike a paddle wheel on a river steamship.

In 2007 the company took a first big step, submerging prototype turbines beneath a ship in Cobscook Bay. Three years later, after tweaking the design, they conducted a pilot test of the full system of turbines and a generator. Then, in 2012, after more changes and a ceremony on the Deep Cove dock complete with speeches by company officials and a representative of the U.S. Department of Energy, ORPC connected the turbine-powered generator through cables into the grid in a deal with Bangor Hydro Electric. For the first time in North America, energy from the ocean powered people's homes. The endeavor was still a modest attempt, with only enough kilowatts to feed twenty-five households. And it lasted just two years, as the generator became waterlogged and would not function efficiently. But the Cobscook Bay test proved in practice that the theory could work.

The turbines' time in operation may come to mark a moment more monumental than the limited fanfare it received in Eastport and among the tidal-power community, but when I walked into ORPC's Eastport office on a cloudy Friday a year after the project ended, the place appeared more like a museum than something moving forward. Desks had been pushed to one side, and one of them was stacked with black phones, no longer needed, as the busy days of action had hit a lull. ORPC had withdrawn to Portland to see what lessons could be learned and to focus its efforts on other projects, including an electricity-generation system along the Kvichak River in Alaska, in which submerged turbines would deliver power to remote communities.

I had come to the fallow ORPC office with Bob Lewis, a local resident who had been hired years before as an on-the-ground jack-of-all-trades for ORPC. Bob has a thick white beard and talks with the no-nonsense seaworthiness of someone who has long lived in a close community bound on all sides by the sea. "Locally, I get asked all the time, 'Has ORPC gone belly up?'" Bob told me. "Have they pulled up stakes?" Though they had, for the time being at least, Bob was quick to explain to me and others around town that ORPC would be back. "Now we're in that normal evaluate-and-redesign-team stage," he said.

So I was left to ponder relics in Eastport. On the back office wall, three versions of early turbine blades hung as an exhibit that captured the individual initiative of the project. The blades each evoked the spin of the seas, and the earliest version was made of a black composite material. A second was natural wood color, and the third, used in the functioning pilot period under Cobscook, was painted white. Two of the blades had been built in 2008 at a local boat school.

The individual pieces of the system were so simple it struck me as a symbol of the work to be done: capturing currents moving across earth is not so much a feat of technical innovation but one of altering basic designs to handle the power of water, wind, and sun. Fossil-fuel extraction requires ever-more chemicals to split deeply buried rock or bigger machines to break open the earth. But the work off the shore of Maine was more about harnessing with elegance the force of nature.

Bob and I talked in the office for an hour, then climbed into his car for a drive out to see where the turbines had been anchored in Cobscook Bay. He detoured first through the old Quoddy housing site, and Bob described growing up in Eastport in the 1950s, when the promise of that project still rang around town. "Quoddy has become a monument to disappointment," Bob said. He predicted that ORPC would be back at work in Eastport, testing new designs in Western Passage, by the next spring. But he took the long view of the work of ORPC and others hoping to harness the tides, and in his perspective echoed the dreams of the 1930s. "I see this as multigenerational. We're laying the foundation," Bob said. "The energy of the water will be tapped. Who actually does it long term, well, history will tell us."

On the way to Cobscook Bay, we stopped at an open lot surrounded by a grass field, where ORPC had put the pieces of its underwater system after the pilot project ended. The assembled turbines lay on their sides, wrapped in tattered blue tarp. Tumbled stacks of white blades rested nearby, as did a carefully milled steel cylinder, which had helped deliver the torque of the turbines to the underwater generator. Next to that sat a towering spool of black electrical cable and a box with a steel frame covered with a messy pile of marine rope.

Bob explained that engineers in ORPC offices were improving the company's TideGen 2.0, the next iteration of the turbine-generator design: changing bearings to produce less friction, for example, and sealing the generator differently to keep water out. He expected the company to be back in Eastport with a new version by 2017. "All this stuff has to be integrated, and there's a learning curve to that," he said.

Even at this exploratory level, Bob said, cost competitiveness was a big challenge. The overall recession that deepened just after ORPC began work in Eastport hurt early fund-raising efforts. And cheaper oil and gas coming with the fracking boom made it tough to convince people to invest in an emerging system that wouldn't pay dividends for years. "It's not a quick turn of your buck," Bob said. "It's about getting people willing to take risks again."

We passed the dock at Deep Cove and stopped for a moment to look out at the blue-black surface of Cobscook Bay. A mile offshore four white buoys still marked the permitted location of the turbine pilot project. But Bob said the place with bigger potential was on the other side of the island, so we made the quick trip back to the center of Eastport, then along a residential street to an overlook of Western Passage. Bob referred to the island as a place that would one day have historical significance, predicting that Eastport would be heralded as a Kittyhawk of innovation in the sea.

If such projections are to come true, much will depend on what happens in the waters that Bob and I faced that afternoon on the east side of the island. Western Passage is only a mile or two across, and through it runs much of the tidal surge traveling between Passamaquoddy and Cobscook Bays. At the south end Old Sow, as it's called, is the largest natural whirlpool in North America, its spinning currents sucking toward the center of nearly an acre on the passage's surface. Out in the center the ebb and flow of more than twenty feet of water rising and falling offers special potential. ORPC aims one day to put a refined system in the heart of the passage, which at points is four hundred feet deep, to test the limits of the technology.

"We want to go right there," Bob told me, pointing north of Old Sow. "That's where the mother lode of the resource is."

The U.S. Department of Energy has tallied the greater potential for such power, measuring tidal flows around the world. Most are less robust than around Eastport, but some are still strong enough to bring energy onshore. Other companies are testing tidal turbine technology in Scotland, Australia, and elsewhere. The DOE predicts that, if technology were perfected and systems installed, as many as ten million households in the United States could run on tidal energy. Such systems could power isolated outposts or feed regional grids, just as wind farms on the windy plains or sun in barren desert can augment power systems. If more robust batteries and transmission lines can store electricity and deliver it to distant markets, ultimately the power of the sea could travel far from the source. As special as Eastport was for Bob, and ORPC, he knew it was not so unique and the lessons learned there could be put to work in coastal communities elsewhere. "It's a big world out there," Bob said, "that has lots of little places like this."

❊

The next day, at 8:08 a.m., captain Bob Peacock and four other men were untying *Zeporah*, a pilot boat, at the dock in Deep Cove. They were there at precisely that time because it was four hours before 12:08 p.m., the ideal target time by which Peacock, who has piloted more than a thousand cargo ships into Eastport, hoped to maneuver *Star Lima*, a 670-foot-long ship flying the Norwegian flag, alongside the pier in the commercial port at the south end of Moose Island. The ship would pick up thousands of pounds of wood pulp and sail the next day for Savona, Italy.

I didn't know any of this at 8:05 a.m., when I wandered aimlessly down to Deep Cove and out onto the dock to look at the sea and saw the crew climbing aboard *Zeporah*. I had been in touch with Peacock by email before arriving in town, but we'd made no plans. When I saw the action around *Zeporah*, with a banner on the bridge that read "PILOT BOAT," I walked down a steep ramp and introduced myself. "Boy did you show up on the right day at the right time," Peacock said. "Want to go for a ride?"

Two of the men were students at the Maine Maritime Academy, and they'd come to get some on-boat experience with Peacock. The other, James Smith, was a local fisherman who held down other seasonal jobs, including running the *Zeporah* with Peacock whenever a ship steamed into the Bay of Fundy bound for Eastport. Soon we were on the pilot deck of the thirty-two-foot powerboat. Smith steered the boat out of the cove and turned south, angling around the southern tip of Moose Island. The city of Eastport and the brick buildings of Water Street appeared to our west, and Smith headed across a stretch of water called Friar Roads toward Campobello Island. He picked up the radio. "Fundy Traffic," he said, "this is Eastport pilot boat. We are entering Canadian waters."

Peacock, who would be climbing aboard *Star Lima* with the two cadets to steer it into Eastport, sat behind Smith. As we neared Campobello Island, he looked at the ebb tide rushing past Windmill Point. The water travels four knots per hour in one direction and then, just past the point, three knots per hour in the other. That morning, as the tide rushed out to sea, the water level around Eastport and in the Friar Roads would drop more than eighteen feet. "We plan everything by the tides," Peacock told me.

Smith steered *Zeporah* past the East Quoddy Lighthouse and into the open water of the Bay of Fundy. Nova Scotia lay fifteen miles east. A mile or two south *Star Lima* was steaming at twelve knots per hour toward a rendezvous point in the deep of the bay. Peacock grabbed the radio and told the *Star Lima* captain to hold his course. "We are dead ahead of you," Peacock said.

*Star Lima* slowed to eight and a half knots, and Smith swung *Zeporah* 180 degrees to come up alongside the ship, heading in the same direction. The current danced on the surface and deep below, and little *Zeporah* rocked sharply from port to starboard and back, like a plastic toy in a tub. Smith worked the throttles as Peacock and the cadets went to the starboard side, where they would climb onto the Jacob's Ladder, a rope-and-rung access hanging down the steel side of the ship. They made the perilous scramble up several rungs, then stepped onto a steel ladder angling up to the deck, as Smith kept *Zeporah* dead even against the ship more than forty times its size. Smith turned his words to me, while keeping his eyes fixed where the starboard rail bobbed inches from *Star Lima*'s hull. "When I first started doing this," he said, "my knees would shake."

My stomach churned and mind bent at the power of the sea, lifting both boats in unison on its swells. The power in the open waters of the Bay of Fundy came from not only the tide but the waves of water moving on the surface. The DOE estimates that wave power, too, could be a robust resource, providing three times the energy of the tides, if harnessed along the coasts of the United States.

As I rode in *Zeporah* alongside *Star Lima*, I thought of my movement around the world, during the years of my walks but also decades of adult life, aboard planes and ships, in trains and cars, always burning fossil fuel to get from one place to the next. The *Star Lima* was just one of thousands of ships sailing the seas morning, noon, and night, each burning thousands of gallons of oil to deliver from one part of the planet to another comforts and conveniences that modern society has come to see as essential. I was hopeful about the power of tides and waves, but had humans already gone too far, building a world of global need that could not be sustained any other way? Even if enough renewable, noncarbon fuel could be captured, ships and other moving vehicles would need to convert to run on electricity rather than oil, and that would require the engineering of electric engines and vast increases of battery-storage technology and transmission capacity.

With Peacock and the two cadets safely aboard *Star Lima*, Smith steered *Zeporah* away, pushing the throttle forward to gain speed for the journey back around the East Quoddy Lighthouse and into the harbor. On its way to the deep-water port, *Star Lima* crossed the waters of Friar Roads and passed within a quarter mile of downtown Eastport. Time collided, with the ship looming larger than a city block of brick buildings that had risen as a center of progress two centuries before.

*Star Lima* arrived in port at 10:30 a.m. but idled offshore for an hour, burning more fuel to stay in place, as it waited for the tide to allow it to dock. At 11:52, sixteen minutes before dead low tide, and one hour and eight minutes before water would begin rushing back toward land, four deckhands on *Star Lima* tossed long lines down to others waiting on the dock, and the ship drew snug against the pier. The Norwegian flag on the deck whipped in the wind as the ship waited to receive its load.

By 1:00 p.m. that afternoon, the tide was gaining momentum on its landward journey, and I followed a foot trail onto a peninsula just north of Deep Cove that is home to Shackford Head State Park. The woods were thick again after earlier logging, and warblers and thrushes darted between trees as I walked. I arrived at a rocky headland more than a hundred feet above Cobscook Bay. To the south *Star Lima* was still at dock. It would take on its load of wood pulp during the afternoon and depart on the next morning's high tide.

The white buoys marking the site of the ORPC pilot project bobbed at their anchors in the bay northwest of where I stood. The local scallop boats, which had been working the bay two days earlier, were back in harbor. The only sound was of the wind, arriving from the north. On the distant shore a farmhouse and fields sat among more forest, and the scene could have been one from long ago.

I knew it would be years, if not decades, before large-scale power from tides would arrive onshore. But wind- and solar-power systems were already far more advanced, so I was happy to be heading next to the Panhandle of Texas, where wind energy was being converted to electricity at a grand scale, proving that with investment and effort the industrial world can be connected to the existing energy of the earth.

Before turning toward home on that last afternoon above Cobscook Bay, I sat on a rock and waited. A seagull soared overhead, calling in the cool gray, and then it was gone around the point, and I was alone again with the distant white buoys. I watched for thirty minutes, as the sea accelerated and the buoys leaned upstream with the incoming tide. It would rise sixteen more feet before dinner.

# Chapter 5
# Turning Time

Across the state of Texas, the administrative anchor of local government tends to be each county's courthouse, and often the buildings are ornate and imposing, meant for the occasion of their office. That is certainly true in Armstrong County, a dry square of earth in the high plains of the Panhandle that measures roughly thirty miles north to south and thirty miles east to west. The southern half of the county is carved by the wide gorge of Palo Duro Canyon, a sharp landscape that offers little comfort for those concerned with survival. Armstrong County is otherwise mostly treeless and flat, with pasture fenced for cattle or fields cultivated for cotton. In the city of Claude, the county seat, the bright brick courthouse, built in 1912 in a classical revival style, still stands four stories tall.

Most days the main action is outside the big building, as tractor trailers and pickup trucks motor through town on the four lanes of Highway 287. The road follows the route of the Fort Worth and Denver City Railway, built in the 1880s, a decade after the U.S. Army forced out Native Americans and settlers began staking tents on the parched plains. Just before nine on a Monday morning in March 2017, a Powder River coal train also rumbled past, on tracks now operated by Union Pacific. The coal train's hundred cars, each with a black cap of the crushed rock, were bound from the Wyoming basin toward power plants lighting the suburban sprawl and skyscraper alleys of Dallas.

But on that morning in Claude, population 1,180, there was action inside the Armstrong County Courthouse too, as twenty-one people crowded beneath the high ceiling in a corner office on the ground floor for a particularly compelling meeting of the commissioners' court. The commissioners would soon hear routine business—about a bridge repair, the sheriff's fee report, and some options for county insurance. But they would give their attention first to Scott Sherwood, a lawyer who had driven from his home in the county just north to represent two businessmen hoping to build a wind farm.

Judge Hugh Reed, a former marine with fine, graying hair, sat at the head of a long table in the center of the office, and on the wall behind him hung a painting depicting a scene from the battle of the Alamo. Three county commissioners joined Reed at the table, and onlookers took seats against the wall and alongside desks. One man, a local pastor, offered me a swivel seat just inside the doorway. I planned that afternoon to take my first steps in neighboring Carson County among cotton and wheat fields already home to huge white towers with blades churning in the sky. Scott Sherwood, whom I'd contacted from afar while planning my route, had suggested I join him first at the Armstrong County Courthouse for the morning meeting.

Reed asked the pastor, who wore jeans and a collared shirt and remained standing by the door, to say a prayer to open the meeting. The pastor first reported some sad news, which explained the absence of a fourth commissioner. That commissioner's grandson had been in an accident while riding just days before. "The horse rolled over on him," the pastor said. "They were on broken ground, and they don't know what happened." The seventeen-year-old had suffered a broken pelvis and brain damage. The pastor bowed his head. "God, we ask that you heal him on your time," he said. After a pause he continued with words for all those in the room: "God, we ask you to bring rain for this land and the farmers and ranchers that work it."

With that, Reed turned to Sherwood, who knew many of the people in the room. Sherwood grew up in the Panhandle and played football at West Point before returning to practice law. He told the commissioners the wind farm was a good idea. The project, called Armstrong County Wind, was spearheaded by two local businessmen, Paul Fields and Tylan Shelton, and it aimed to plant 118 towers into the prairie to capture currents and send the power into the grid. Sherwood asked the commissioners to consider carefully a request, which would come that morning from Fields and Shelton, to create a special business zone and offer a tax break for the wind farm. "We're trying to lure them here to make sure they do it here," Sherwood told the commissioners. He encouraged them to be greedy enough to welcome Armstrong County Wind, but not so greedy as to deny the tax-break request and risk chasing the project away. "It's the classic situation," Sherwood told the men around the table. "Pigs get fat, and hogs get slaughtered."

The commissioners knew their county could use some fattening up. There was none of the oil, gas, and minerals beneath the soil of Armstrong County that had brought riches to other parts of Texas, even neighboring counties. And they knew profiting from the wind—hot gusts of which had long cooked crops and scattered dust—was no longer a far-fetched idea. During the previous decade dozens of industrial wind farms had taken root in Texas, particularly in the Permian Basin a couple hundred miles south and more recently in the Panhandle. The state, known for more than a century as a leader in drilling oil, had quickly become the largest single producer of wind energy in the United States.

Sherwood had some credibility with the commissioners gathered around the table. As county attorney of neighboring Carson County, he'd helped orchestrate the building of several wind farms there, with hundreds of turbines turning morning, noon, and night, bringing leasing fees and royalties to the landowners. He suggested Armstrong County, where another developer was considering a second wind farm, get in on the action. "You're not losing anything," he told the commissioners, referencing the potential tax break, "because you don't have it right now."

The local deliberation in the corner office was exactly the kind of thing that seemed so crucial after my walk along the tidal coast of Maine: a community teaming up with businesses hoping to make money from renewable energy. The technology exists to allow much more noncarbon fuel—especially from wind and solar energy—to power much more of the electricity grid than it already does. But those fuels need to be made a cultural and commercial priority. Continued momentum for such a shift would need to include action in congressional halls and Wall Street boardrooms. Smaller steps would occur in places such as Armstrong County's brick courthouse, as communities choose whether or not to invest in the shift at ground level.

After Sherwood made his case, he turned to Shelton, who had been sitting against the wall, shifting in his seat, as though trying to stay calm in the face of a big decision. Shelton has roots in Armstrong County but lives in Amarillo, thirty minutes northwest, and he had the benefit of local ties with a big-city shine. He had a thick gray goatee and wore a sport coat over his open-collar shirt.

He planted his cowboy boots firmly on the ground and leaned forward as he told the commissioners a feel-good story. He had already instigated the building of one large wind farm in the region, and he remembered a conversation he had with Fields, the local farmer and businessman, who was part of the earlier project and had five turbines built on his land. After construction was done and royalty checks started to arrive, Fields called Shelton and said, "That was easy. Let's do another one."

With a wide smile Shelton told the commissioners that it wasn't so simple as that, but he had already lined up twenty-five Armstrong County landowners, who control twenty thousand acres, for this new farm. Armstrong County Wind had gone through months of regulatory review, Shelton said, and received permits from U.S. Fish and Wildlife, the Federal Aviation Administration, and other agencies. It had collected eighteen months of wind data on the prospective site of the wind farm. "It's showing a very good wind regime, which we all know," Shelton said.

The project design called for two-megawatt turbines, and the hard work, Shelton said, lay ahead. Shelton and Fields would attempt to sell the project and its permits to a major energy corporation, such as E.ON Climate & Renewables North America or Pattern Energy, which operate farms in the Panhandle. Such a buyer would finance the $325 million cost to build the farm. That, Shelton told the commissioners, would mean 150 jobs during the twelve months of construction, with 8 to 10 permanent jobs afterward.

Shelton said that Armstrong County Wind, with its land leases and federal permits, was already worth $7 or $8 million to a potential bidder. To tempt those bigger corporate investors, Shelton said, the project would need special permission to operate and a tax break from the county. Armstrong County Wind would offer the county a "PIL," or payment in lieu of taxes, of nearly $500,000 per year during that decade. "What we're humbly asking for is that the court consider the reinvestment zone and a tax abatement at $1,800 a megawatt per year, for ten years," Shelton said.

One of the younger men at the table, the newest commissioner on the court, joked that others would have to correct him if he didn't understand how the process worked. But making the decision was, in any case, a casual affair, with Reed soliciting questions for Shelton from the commissioners and the crowd. The younger commissioner asked Shelton if there was any guarantee locals would be hired during the construction phase. Shelton pointed out that some workers would come from elsewhere. But he told the court he would encourage any eventual developer to hire in and around Claude where possible. "I'll always be a local advocate," Shelton said.

The sheriff stepped through the doorway to say she'd like to know if there would be drug and other background checks when hiring workers, as she'd seen other counties struggle to police the boom-time community that rolls in during construction. Shelton assured her that any construction company would carefully screen its workers.

A red-haired man with broad shoulders who owned a trucking company asked Shelton when construction might start, as he would like to get a piece of the action. Shelton said that if all went well construction would begin late in 2017, with the wind farm up and running in 2018. Another man, involved in the local volunteer fire crew, asked if there might be extra concrete from the wind-turbine foundations that the county could have to build a helicopter pad near the firehouse. Shelton said he imagined there might be.

And that was all: focused questions of a local nature that had more to do with the terrain outside the courthouse than the wider world and the challenges it faced. There was not, in the room before or after Shelton's pitch, any talk of climate change, broader

renewable-energy policy, or the urgent need to find a new way to
fuel the industrial system.

The passing coal train, in other words, was not seen by those
in the courthouse as a problem to be removed, as much as the
construction of a wind farm was an opportunity to keep moving
forward. In Texas, at least, this was key to the new system getting
built: communities perceiving immediate gain—of jobs, royalties,
or a general sense of local progress. It was clear, as the commissioners considered the offer, that if society is ever to develop more
renewable fuel, financial mechanisms must encourage action:
tax breaks to make it more affordable to build and surcharges on
those who burn carbon, as a deterrent to following the easy path.

After a few minutes of conversation, the commissioners asked
others in the room to leave, and they voted unanimously in closed
session to create a redevelopment zone that would allow for a
wind farm and to grant the ten-year tax abatement to Armstrong
County Wind. The room filled again, and Judge Reed announced
the decision.

"Hopefully," Shelton said, "it's literally a windfall for the
county."

The commissioners turned to other business, and many of the
people who'd taken seats around the room stood to leave. It was
after ten o'clock, the temperature approaching eighty degrees. On
the courthouse lawn, steady gusts were rising too. Doves sat on
branches in the shade of a tree, their soft coos drifting. As the
corner-office crowd shuffled toward the bright and empty day,
one person said, "Everybody's going to be praising the wind now."

※

Wind is the sun's energy in motion. Solar rays heat air at varying
rates, depending on the angle of the sun and features on the earth's
surface. As warmer air rises, cooler air rushes in to take its place,
creating currents. The rotation of the earth, and its seasonal tilts

toward and away from the sun, creates global wind patterns, with doldrums along the equator. In the Northern Hemisphere, trade winds blow south, and westerlies angle north, and, near the pole, easterlies head south again. Within these zones oceans, lakes, mountains and other surface features create localized trends, and in certain places a combination of conditions causes strong and regular wind.

In the United States there are especially windy pockets in California, the interior of the Pacific Northwest, and along the Appalachian Mountains. But the most prominent winds blow across the northern plains and in west Texas, where the strongest of all rise and fall from day to night, as the flat earth warms and cools. Wind maps of the United States, color-coded documents indicating the strength and frequency of moving air, are darkest across a swath of land that includes the Texas Panhandle.

The region has been home to people for fourteen thousand years, with one community more than eight hundred years ago growing squash and beans along a river. But cracked earth most often meant more nomadic living, and Comanches had controlled the Panhandle until the army drove them out during the Red River War of the 1870s. The settlers who followed cursed the constant companionship of the whipping wind, but by the late 1800s they were building wooden windmills to draw up water from deep underground, allowing for farming and the raising of livestock. By 1900 most windmills were built of metal, with blades fifteen feet across. By the 1930s, a peak for rural wind turbines, the blades were turning not only to pump water but also to convert power into electricity for small appliances. "If the wind blowed, we got to listen to the radio that night," Delbert Trew, a rancher now in his eighties, told me of those days. But by the 1940s rural electric grids began delivering a fossil fuel–fired world onto the plains, and the old towers were left to fall to the ground. Some are still standing. Their blades turn in heavy gusts, testimony to days of purpose.

In the 1970s, during the global oil crisis, the Alternative Energy Institute took root at West Texas State, in the city of Canyon. Its staff studied early wind-turbine technology and advocated for the development of wind farms to provide electricity for the modern grid. The motivation then was concern about *peak oil*, the point at which there won't be enough petroleum to burn. It has only been in recent decades, with worry about carbon emissions changing the atmosphere, that wind-energy businesses began to grow in Texas and around the world.

The National Renewable Energy Laboratory, a division of the U.S. Department of Energy, estimates that the so-called technical wind resource—the amount that could be captured if technology were in place—is more than thirty-three thousand terawatt hours: enough to provide more than eight times the electricity consumed nationwide. Even a fraction of that could go a long way, and in the past twenty years wind power has boomed at particular points across the country. The U.S. Energy Information Administration announced that wind power outpaced hydroelectric generation in March 2017, the month I walked into the Panhandle, as the largest single source of renewable energy in the United States, generating 8 percent of all electricity. It was the first time since the 1930s, the EIA stated, that a renewable source other than hydro had been most dominant. The benchmark capped a decade of particularly fast growth for wind, with eleven states producing more than 10 percent of their energy from industrial-scale farms. Those have blossomed in California, long a leader in renewable-energy infrastructure, and across the plains states, with Iowa, Kansas, South Dakota, and, increasingly, Oklahoma building big networks of towers and transmission.

But energy infrastructure and landowner philosophy have helped Texas become far and away the largest wind-energy producer. Most electricity grids in the United States are regional, overlaying many states in whole or in part, which makes for difficult policy and planning. Texas is largely powered by a grid managed by the Electric Reliability Council of Texas, located entirely within the state. ERCOT, as it's called, is overseen by the Texas legislature, which has promoted renewable-energy production and power transmission–line investment. Texas landowners, who long have operated under a philosophy of law called *rule of capture*—which gives private-property owners claim to resources, such as oil and water deep underground—were preconditioned to the idea of harnessing their wind resource. So some of the world's largest wind farms have been built in Texas, mostly in the Permian Basin and along the Gulf coast, both of which lay within ERCOT territory. A rugged physical boundary called the *cap rock* had separated the Panhandle from the ERCOT grid for decades. But the construction of a powerful transmission line connecting the Panhandle to ERCOT was completed in 2013, opening the market for Panhandle wind power to follow the coal trains toward Dallas, Austin, San Antonio, and Houston.

In the years prior to my visit, six major wind farms had been built in Carson County, just north of Armstrong. That growth had also accelerated because wind companies didn't want to miss out on federal tax credits. A program begun in the early 1990s to stimulate the wind industry was set to expire in 2019, with the credits phased out in the years before.

The most abundant wind in the state blows in the Panhandle. As Ken Starcher, an engineer who has worked at the Alternative Energy Institute for decades, had told me, "We've got high air moving fast." On that Monday in March, I would begin moving among hundreds of wind towers, each looming as high as a city skyscraper. The blades turned in time with the rising and falling wind, and the blinking red lights atop each turbine were already as much a part of the view at night as the stars overhead.

A few hours after the Armstrong County commissioners had voted on the tax break, I drove a small rental car into the town of

Groom, about twenty miles northeast of Claude. The city sits at the southeastern edge of Carson County, and the streets of Groom, population 574, are arranged in a grid, as is typical of small towns in the plains. I pulled to a stop at the intersection where County Road 295 meets Business Route 40, one of the rare stretches of working road that still follows the path of historic Route 66. It was early afternoon, and hot, and there was no traffic to speak of in any direction, except for a giant, lumbering farm machine that was approaching the intersection from the east on Highway 40. I was planning to turn that direction and continue three blocks to the Attebury Grain Elevator. Scott Sherwood had arranged for me to leave the rental car behind the elevator during my week of walking. I planned to cover twelve miles on foot that afternoon, the beginning of a forty-five-mile arc that would take me through three of Carson County's largest wind farms. I would stop for the night in a cotton field owned by a local farmer who had agreed to let me sleep there. The cotton had recently been harvested, but crucial for my interest: the field was also home to three wind turbines.

All this was spinning in my head as I sat in my rental car at the stop sign at the intersection of 295 and 40 and watched the farm machine rumble in my direction. I assumed it would continue straight through the intersection, as it was easily rolling twenty miles per hour. As it got closer, I saw that the machine's single front tire was five or six feet tall and several feet wide. The driver, whom I could not see, was in an enclosed cab at the center. Two long arms, with nozzles spaced along their lengths and folded against the cab, gave the contraption the look of a steel spider.

Then the big machine did something I did not expect: as the front wheel rolled into the intersection, it began to turn in my direction, though not soon enough that it would make a tight right angle into the oncoming lane. For a moment I figured the driver must have been planning to cut close across the front of my car while angling out into the adjacent field. But the front wheel kept turning. Suddenly the wheel was only a few feet from the hood of my little Hyundai Accent, and the machine rose up behind it like a cresting wave. I remember looking up at the cab—all shadowed glass and anonymity—and verbally screaming, as I shoved the automatic transmission toward reverse and moved my foot to slam on the gas.

When I thought of that moment later, it seemed to extend for several seconds, everything slowing to a crawl, with the chain of events unfolding as though from a script, clear in their purpose. In recent decades the family farming that had moved west during two centuries had become as industrialized as the big cities the farms fed. Fields were cleared to raise a single type of crop or animal, and ever-larger machines were used for everything from seeding to irrigation to harvest. Just as Leif Jellesed had told me in North Dakota, a farmer's main work was in sitting high in the air-conditioned cab and watching dials and pushing buttons and turning joysticks. How could a driver of such a thing, insulated against the earth in every way, possibly be aware of what was happening at ground level, let alone directly in front of him? The system was out of proportion to its surroundings, and all its pieces, including this single big farm machine, operated above the rules of nature. Collisions were inevitable.

I felt in that suspended moment as the machine bore down on me as though I would become a piece of collateral damage in the accelerated life of 2017. As I imagine may have been true of the smashed turtle I encountered on the road in New York or the flattened bird I also saw there, I knew for a split second a sensation of emptiness that arrived as the rubber tire slammed into my little sedan, its steel hood and plastic bumper and low-grade underbody bowing in submission to the creations that now dictate the pace at which life is lived.

I am not sure if in that moment of seemingly slowed impact my

foot actually made it to the gas pedal, punching down and sending gasoline into the engine to fire the pistons in enough time to
get me moving in reverse. Because as soon as that big machine's
front tire crashed into my car, everything was going backward in
a single, vaulting flash. There was a smash as the hard rubber of
the tire moved onto the passenger side of the hood and over the
other side, and the car spit backward like a seed, several feet into
the lane in which I'd been stopped. Pools of coolant and windshield washer fluid spilled as their containers shattered, covering the ground where I'd been idling, leaving a stain of roadkill
already cleared away.

I was in a state of shock, I'm sure, but saw soon enough that the
driver's side of my car had not been crushed, and my legs and arms
and everything, really, was fine. I opened the door and climbed
out, standing on the asphalt. Seth Ruthardt, as I would learn his
name to be, had already parked the huge farm vehicle, which I
would learn is called a Terragator. The young man came lumbering toward me with a bewildered look in his eye, as though he'd
made a crash landing on the surface of the moon. Seth had only
begun driving the machine a week or two before. "I didn't even
see you," he told me. A Terragator weighs ten tons when its cargo
bin is empty. On that afternoon Seth had been carrying a full load
of fertilizer that he planned to spray on a field. He'd rounded the
wide turn of the corner in a machine that weighed more than
twenty tons. "I tried to step on the brakes, but it wouldn't stop,"
Seth said.

We stood there, the two of us, sharing an odd bond of unexpected violence, and waited in the steady wind as first a state
trooper, then the county sheriff, pulled up to the intersection. The
trooper interviewed Seth, then me, about what happened. He had
an air of purpose but also of nonchalance, as though the scene
were normal, people living by the whims of the machines long
ago set in motion.

A flatbed wrecker came next, to haul my rental car away. There was no damage to the Terragator, which Seth would soon drive into the field to finish its work. The driver of the wrecker, named Will, was lean and moved with the fast twitch of a man trying to correct things or at least carry away the inconvenience. He attached a steel cable under my rental car and used a hydraulic winch to pull it onto his flatbed. He planned to drive it to a storage lot in the city of Panhandle, where I was going to end my walk through the wind farms a week later. Will offered to give me a ride there, but I asked instead if he could drop me on the other edge of Groom, where there was a simple country motel. Will shrugged and drove the two miles along tree-lined streets, then turned into the motel's parking lot.

During the two-plus years of my ongoing odyssey on foot, I had occasionally questioned the wisdom of walking. Did I really need to put myself closer to the earth to sense our relationship with nature? But time and again, I'd known its value: moving under my own effort, I traveled beyond the overinsulation of modern life, and in the sharpest moments I saw the toll of that separation from nature more clearly. The crash at the intersection in Groom confirmed my instincts to move at ground level, and it gave me courage for the journey ahead. I thanked Will for the ride, climbed onto the flatbed, and took my backpack and walking stick from the car.

Inside the motel office the young woman behind the counter watched wide-eyed as Will drove my wreck away. I wanted a good night's sleep and booked a room. The woman asked where I would go the next day. I told the woman I would head toward Panhandle. She wrinkled her brow, as though measuring the open prairie that lay between, and asked how I would get another car for the journey. I thought to explain but said, "I'll just walk."

*

I had walked away from the motel the day before, and in the dead of that next night the near-full moon brought a dusky glow to the sky above my tent. I shifted in my sleeping bag, the chunky clods of dry earth crumbling beneath my weight. Down low in the field, among the harvested cotton plants, shadows of twisted branches added depth to the darkness. Across the broad horizon, hundreds of red lights set atop wind towers blinked to different beats, their on-and-off signals seeming to form an organism poised above the plains. A few hundred feet east and west of my tent, towers rose 240 feet into the sky. I sensed their presence in the sweep of the three long blades turning atop each tower in the wind. The blades measured more than 300 feet in diameter, and their relentless rotation cut the air with a deep *whomp, whomp, whomp*. The air temperature had dropped to twenty-eight degrees. I wore several light layers of clothing and tried to nestle deeper into my too-thin sleeping bag. The cold of my stillness braced against the energy of the windmills spinning high above.

Wind in west Texas generally rises and falls twice a day, with the strongest gusts coming each afternoon and again in the hours after midnight. Ken, the Alternative Energy Institute wind expert, told me before I started walking that the wind would blow strong through the night, and he was right. The blades overhead turned with such force that it felt in my tent as though the turbines may lift the field into the sky, everything floating into the moon glow, a magic carpet ride on the unseen energy. Each set of turning blades drove a rotor that accelerated the spin of a shaft atop the wind tower. That shaft was connected to a generator that created electricity from the wind's power. The electrical currents traveled through cables buried in the field that connect to power lines running alongside the road. Those continued to an electrical substation, which shipped the electricity along high-powered transmission lines toward the urban corridors, where millions of people would immediately consume it.

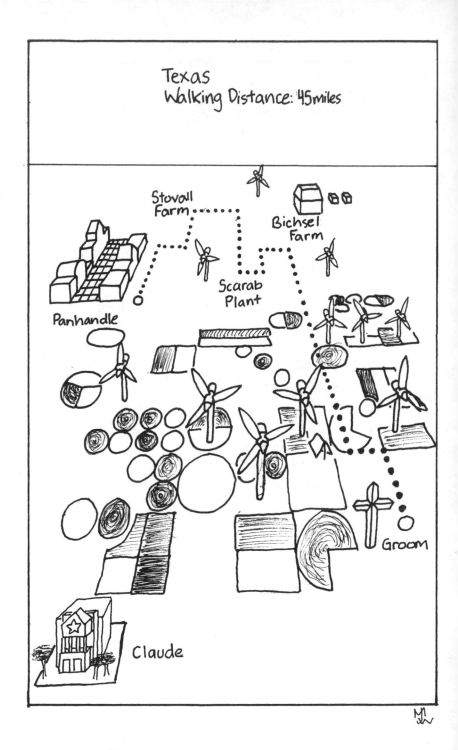

Texas
Walking Distance: 45miles

Stovall Farm

Bichsel Farm

Scarab Plant

Panhandle

Groom

Claude

The three towers on the half section of land where I slept were part of the 118-tower Grandview Wind Farm, and such an operation required deep resources. The farm was jointly owned by E.ON Climate & Renewables North America, a subsidiary of the world's largest private utility; and GE Energy Financial Services. General Electric turbines sat atop each tower, and the construction of the farm was said to cost more than $400 million. Such investment was clearly aimed at harnessing wind that could be used now, but also to learn lessons for future projects. Said Patrick Woodson, an E.ON executive, when the Grandview Wind Farm began producing power in 2015: "The outstanding nature of the winds here will test our technology and operating approaches."

The wind blows harder at the height of the towers than it does at ground level. The ideal speed to drive the blades ranges from twenty to twenty-five miles per hour: strong enough to provide good energy without putting too much pressure on the turbine and generator. I did not know how hard the wind was blowing in that dead of night as I lay on the ground in my tent, only that it was constant and cold.

I had arrived in the field the evening before, after the twelve-mile walk from the motel in Groom. My route had traced mile-long sections of dirt county road, first north, then west, then north again, and the terrain along the way had been flat and open, mostly brown fields awaiting the planting of crops in the coming weeks. The rows of wind turbines began immediately at the north edge of Groom, and by the time I'd walked an hour or two I could spin slowly in a spot and count hundreds in the distance in every direction. But the towers were several hundred yards or more back from the road, and I did not get a chance to stand up close to one.

During one stretch of walking in a particularly punishing part of the day, hot gusts coming constantly, I saw a person on horseback up the road. The horse was coming in my direction, and after fifteen minutes the rider and I passed each other only a few feet apart. The young woman in the saddle said she'd been scared to see me out there on foot, as other people are rare, and those who pass do so in trucks or tractors.

I saw only a few vehicles all afternoon, and in the mile before arriving at the cotton field to camp for the night, a passing pickup had slowed to a stop. Two adolescent boys sat on the bench seat alongside a burly man, their father, who introduced himself as Hop Britten. He laughed as I stood there with my pack on my back. An older son of his had hiked the Appalachian Trail, and Hop told me his boy used to train by hiking these same dirt farm roads. He wondered what I was up to, and, after I explained, he raved about how wind had helped local farmers. He cited as an example a farmer who might have five or six towers on his land. Each acre with a tower could generate roughly $9,000 a year for the farmer. That same acre could bring $800 a year in crops. Yet the towers took up so little space, that the farmer could still raise corn, sorghum, or cotton around the towers. The wind, in other words, functioned as another crop, bringing diversity to the operation.

Hop had pulled to a stop in an intersection, and during the first few minutes we talked no other vehicles approached. But suddenly I saw a huge machine motoring our direction from the east, and as it got closer I saw it was a Terragator, possibly even the same one that had hit my car in Groom the day before. I told Hop and his kids about the collision, and they laughed, seeing that I'd lived to tell the tale. Hop looked to the west, the direction the Terragator had gone after it sped past us. "That thing is out here hunting you," he said, smiling.

Hop encouraged me to walk a mile east to his parents' farm, where I could have a spare room for the night. But I told him I was eager, after my hot afternoon of walking, to spend more leisurely time with the wind. Hop and his boys sped south toward Groom, and I slowly made my way out into the cotton field. At home in New Hampshire, I cycle regularly and go to a gym for fitness. But it is a different thing to walk all afternoon, a week's worth of supplies in a pack. And on the first days of each of my walks, I tired quickly, with aches in shoulders and hips, thighs and feet. Making my way into the cotton field, I had a burning pain in the sole of my left foot. I limped along and finally dropped my pack on a short set of metal stairs at the base of the first tower.

There had been no warning signs at the edge of the field, no announcements of dangerous chemicals or commands to keep away from the tower. The tower was bright white, and at the base, which measured roughly twenty feet across, giant bolts connected the tower to a concrete pad. A sign on the door leading from the stairs into the tower announced it was private property, and it should not be disturbed. But I was free to idle beneath the blades.

After sitting and drinking water to regain strength, I stood and tilted my head back, looking straight up at the blades turning overhead. Their length overwhelmed, and each rotated with a velocity that seemed to increase before rising up again in its arc. The *whomp, whomp, whomp* of the rotation echoed into the air, but also into my chest, as the power was profound.

The evening air was still warm, and the field held a certain promise, as the brittle stalks of cotton waited for another season, and the wind worked the turbine overhead. Songbirds swept past near ground level, settling only briefly in the breeze before lifting again. I spun beneath the tower, looking out at the hundreds of other white towers trailing toward the horizon. Their presence no doubt changed the character of the landscape. The machine overhead dominated, but it operated with predictable grace. I could not fathom the amount of power turning above me in the single turbine, all driven by the existing wind. Its constant rotations offered the kind of hope I had been seeking. According to the leading wind-energy trade association, more than fifty thousand wind turbines stand across the United States, with more than twenty thousand of those in Texas. Will wind farms one day cover the plains and coastlines?

A frequent critique of wind power is that it is not constant and can serve only when the wind blows, and then only in windy places. But with more robust transmission lines and better battery storage, much more of that wind power could be captured. Storage has been elusive so far, but the International Energy Agency described the five years before my walk, during which lithium-ion batteries became the dominant technology for utility-scale installations, as a "period of ferment" that may lead to more innovation. Billionaire entrepreneur Elon Musk built a huge battery-manufacturing plant in Nevada, racing others in the hopes of cracking the electricity-storage riddle.

As I stood beneath the spinning blades, I arced back again, partly to stretch tired muscles but also to face the force of the air overhead. I was tempted for several minutes to make camp in that exact spot. But the *whomp, whomp, whomp*, so promising in the energy it conveyed, felt too fierce on an individual level. I struck a compromise, pitching my tent a few hundred feet farther west, between the first tower and a second. I cooked a simple supper on my propane stove and crawled in to sleep.

When I woke at seven, the walls of my tent hung loose at their stakes. The towers overhead were silent. Ken had told me that the diurnal wind patterns decreased near daybreak. I lay in my sleeping bag for thirty minutes, and more, waiting as the first rays brightened the day and moved the birds to song. Conservationists have long criticized the threat wind towers pose to birds. While exact counts vary widely and are hard to gauge, the American Bird Conservancy estimates that hundreds of thousands of birds and bats are killed each year by turbines. Do the threats of towers to wildlife outweigh the benefits of carbon-free power in an age of climate change?

At ground level among the cotton plants, as I lay in my tent, only the songbirds moved quickly. Atop the tower just to the east, a motor clicked and whirred, as computers sensed the lull and turned the blades several degrees toward a predicted shift in the wind. I waited longer, my nose and face exposed to the chill, the temperature still in the twenties. Sure enough, by eight o'clock, with the sun higher in the sky, a breeze was buffeting my tent, and, 240 feet above, rising gusts had set the blades turning into another day.

By nine o'clock I was both rested and restless. After a cup of hot tea, I hurried to pack my tent. My fingers warmed with the work. I shouldered my backpack and walked from the field, turning north for another ten-mile walk toward the town of White Deer. The day before, my steps had honed my attention on smaller details. I studied a small trail that rabbits followed from a roadside ditch into a lot full of hay bales. Later I watched sandhill cranes making the spring migration from Mexico north toward Nebraska's Platte River and on to Canada, passing in shape-shifting flocks high overhead. I did not know if they were a quarter mile, or higher, above, but they were riding currents of air even more powerful than those turning the turbines atop the towers.

I swung my walking stick forward, and my legs and shoulders settled into an easy pace. I passed a pasture, where two grazing horses bolted from their places, then circled back to stare at me, curious. I said, "Good morning," and they bolted again. I stared straight up, sure to see more sandhill cranes. They would soon be arcing high overhead, following their natural course. I kept walking for one mile, and another, ever more wind towers on the horizon. Together they formed an endless pattern, as a tide rising and falling in the sea.

※

The grounds of a low-slung manufacturing plant west of White Deer had the feel of an oasis among the ever-flat fields. Trees

on the property shaded a house on the left and a one-story office entrance. A machine shop was out back. Marvin Urbanczyk,
the eighty-two-year-old man who four decades earlier had started
the company Scarab International, which designs and manufac-
tures giant composting machines, offered me a seat in the small
lobby. The office also served as a taxidermy zoo, with the heads
and in some cases full bodies of dozens of stuffed animals from
North America, Africa, and elsewhere hanging on the walls and
mounted on the floor: a polar bear, a grizzly, a giraffe, a lion, a
leopard, a zebra, a musk ox, and a baboon. Marvin had shot them
all, many on side trips while doing business for the giant Scarab
machines, which have been shipped to customers in Egypt and
China, New Zealand and Brazil. "Any organic material, anything
that's ever lived, we can make fertilizer out of it," Marvin told me
about the composters he designed.

My destination for the day was a family farm five miles north,
but since arriving in Carson County I'd heard about Scarab, and
the cotton field I had camped in the night before was owned
by Marvin. So I stopped in for a visit. Marvin and I wandered
through the machine shop, where workers were milling metal for
the giant composters, rolling contraptions with stout steel legs
that looked not unlike the *Scarabaeus sacer*, the beetle for which
they are named. The beetle survives by pushing a ball of waste as
it gathers mass and then digging itself inside the ball. There it eats
its way out, only to begin the cycle again. Customers use Marvin's
machines, which measure anywhere from sixteen to forty feet
wide, to roll above long rows of compost, lifting the dirt up and
turning it over to accelerate the natural process of decomposition.
Marvin invented the design for the first machine in the 1970s by
converting an old Ford tractor after the owner of a neighboring
cattle operation asked him to devise some use for his bounty of
cow manure.

Marvin led me out the back of the shop to a lot where he keeps combines and other modern tools for working the land all around. Marvin's grandfather had emigrated from Poland to South Texas in 1905 and then west. When Marvin was a boy in Carson County, a big plow on a tractor with a 40-horsepower engine measured twelve feet across. He would sit in an open seat in the biting wind, spending all day just to work a small section of earth. We climbed into the climate-controlled cab of Marvin's newest tractor, which has a 450-horsepower engine. That afternoon Marvin would till rows with a plow sixty feet wide. He would sit in the air-conditioning and listen to the radio while the tractor did its work. On cool mornings he could warm the cab too. "It doesn't have a heated steering wheel, though," Marvin said, with a laugh. "It's a little antiquated."

Marvin and I took a quick turn back through the machine shop, passing a doorway that led directly into Marvin's house. Though his son, Mark, runs the Scarab business, Marvin still keeps an eye on the manufacturing of the composters. He would like to make them better. I wondered if Marvin might be skeptical of the wind boom that has come to this rugged terrain. "Wind in the prairie is a pain in the ass for us," Marvin told me. "It's just an evil that we have to put up with." Marvin and Mark have twenty-five towers on land they own. "This is making something good for us," he said.

When Marvin was a boy, oil wells were drilled on the family land, as some oil sits underneath Carson County. The oil field is not as large as that in Borger, forty miles north. But the wells brought up crude for decades. "Most of the wells drilled when Dad had the land are dry already," Marvin said. "These windmills will be here forever." He knew from the profits he took from the towers on his land and the experience he had as a world-traveling business owner that there was plenty of financial motivation to keep wind development going. "If they weren't making money, all these damn big companies wouldn't be out here," Marvin said.

He led me back to the lobby and said I was welcome to linger among the stuffed animals. I filled my water bottles, and Marvin showed me a safe with a collection of high-powered hunting rifles and ammunition. He was planning a trip the next month to South Africa. He'd do business for Scarab along the way, hoping to sell more composters. But he'd also leave time for hunting. He wanted this time to shoot a hyena, which he'd never done. Then Marvin said he had some work to do, so I sat alone and snacked on a piece of beef jerky I'd been carrying in my pack. I made a few notes, then said good-bye to the receptionist, who sat at a desk not far from the stuffed bears.

I stepped through the front door into hot wind and turned onto the dirt road, happy to have that wind at my back. I soon passed a cotton gin that a cooperative of local farmers use for their crop. A dump truck pulled into the yard, and I jogged to beat traffic while crossing the two lanes of U.S. Highway 60. The road ran in a straight shot from White Deer, just east, to the city of Panhandle, fifteen miles southwest. I would angle west and south toward Panhandle in the next days, but I continued that afternoon due north toward the Bichsel family farm. Tracks lay just beyond the highway, and I had to stop for a passing train. I stared straight ahead at hundreds of boxcars carrying cargo.

The train gone, I stepped across the tracks. The scene as I moved north was marked by broad swaths: brown and green fields below, blue sky above. White wind towers punctuated the terrain as far as I could see. There were so few elements around me I was able to notice what I already knew: the towers to the west were as tall as those I'd camped beneath the night before. They belonged to a newly constructed wind farm, built and managed by Pattern Energy, a multinational corporation with projects in Asia, Europe, and North America. But the towers north measured only 150 feet high. They had been built more than fifteen years before, as part of Llano Estacado Wind Ranch, the first wind farm constructed in the Panhandle. The turbines atop each tower produced less power than the newest machines. But they had been doing their work for more than a decade, and as I approached on foot they kept the wind's cadence with relentless conviction.

The road ahead rose slightly toward the horizon, and with gusts pushing from behind, I felt as though I were gliding. On that second afternoon of walking, my legs strengthened, my shoulders settled against the load of my pack, and my mind sharpened with the task of moving myself across such a vast place, exposed.

The words of Scott Sherwood, the attorney who'd helped me on my way, held deeper resonance when I considered the broader terrain: "Pigs get fat, and hogs get slaughtered." If the goal is to keep the comforts and convenience of an industrial life, that life must be powered by a new fuel system. Some of that innovation may be technical, such as more robust battery storage. Other may involve infrastructure, such as more expansive transmission lines. But it all involves a measured choice: to be ambitious enough to get fat but not so greedy as to end up slaughtered. That re-engineering of ambitions requires policies promoting renewable, noncarbon energy so that money can be made. Embracing such policies will come with a more philosophical reckoning: that it is worth working at the natural scale of these elemental sources.

I walked with the wind for another hour, my stick swinging with each step, and I turned into a long driveway and saw up ahead, beyond a farmhouse and barn, a lone tractor making slow passes. Three white wind towers stood in a line alongside the drive, the last only a few hundred yards from the farmhouse. As I approached, a young man cut behind the barn toward a pen, where he kept calves. Marjorie Bichsel opened the front door of the house and stepped out on the porch. I was about to say hello, when Marjorie said, "Are you hungry?"

She invited me inside, and I sat at the dining table next to the kitchen as she cooked meatloaf, potatoes, and peas. She talked about her childhood in northeastern Oklahoma not long after the Dustbowl, when she and her nine siblings grew up in a house that often didn't have electricity. After high school, her family moved to the Panhandle, and she met Robert Bichsel, who'd been born and raised on the farm where the couple still lived. In the 1990s there had been little talk of wind energy in Carson County, until a local developer pitched the idea of putting towers for the Llano Estacado Wind Ranch on the Bichsels' property. The Bichsels needed help making money to keep the farm. "It has been a blessing," Marjorie said. "In the bad years, when we don't make nothing farming, it has kept us going."

Robert Bichsel soon came in from the tractor, and Andrew, their grandson, from the calves. We ate and talked about the wind boom that finally came to the rest of the county more than a decade after Llano Estacado began. Marjorie was excited about her grandson's studies at a community college in Amarillo, but less so about the prospects for their aging wind towers. A few months earlier the Llano Estacado wind farm had been sold to a company in New Jersey, and the Bichsels recently received notice that their royalties would decrease. The turbines and blades were nearing the end of their lifespan, and it wasn't clear whether the parts would be replaced, or if the wind farm would simply shut down, the towers left to weather in the wind like those built of wood more than a century before.

Marjorie offered dessert, but I was tired, and dusk had come quickly. I wanted to sleep outside and asked if I could pitch my tent in a side yard, surrounded by a wall of short, stout pines, planted to break the wind. After dark I looked south, my view across the plains more expansive on this piece of higher ground. I saw hundreds of blinking red lights, and I knew that for each light three blades were turning. I soon was fast asleep, but awoke with a start sometime in the night. Coyotes in the distance were yipping and yelping, and another coyote not far from my tent answered their call. Had a coyote killed a rabbit or even a calf, celebrating a feast that would carry it further?

⁂

By ten in the morning the temperature had climbed into the eighties, and gusts were moving above twenty miles per hour at ground level. I hoped to cover six quick miles from the Bichsels' farm, west, then south to reach the field office of Pattern Energy. From inside the building, engineers and mechanics responded to data collected by computers to control nearly two hundred turbines that are part of two large wind farms. The visit would offer me an inside view of what seemed, when I was walking, to be a simple process of turning blades through time.

My route headed straight into the wind, and after an hour I set down my pack and slumped against a fence post. A dirt track ran toward a row of Pattern's wind towers. I had just sat down when a pickup truck approached from the north. It slowed to a stop, and I saw a woman in the passenger seat. An older man, lean with white hair, got out on the driver's side. He introduced himself as the owner of the land all around, and I explained I was just taking a rest while passing through. He said I was welcome. When he heard I was writing about wind power, his mood soured. He had six towers on his land, including those just west of where we stood. He had waited until all his neighbors had signed on with Pattern, and then figured he may as well go along. The towers brought decent money, he said, but not as much as Pattern had predicted, and it seemed to him the turbines had more mechanical problems than they should. He pointed toward the horizon, a mile or two away, where a repair crane stood next to a tower. "What is happening there?" he said. Two years earlier, just before construction crews arrived to put up the towers, he drove out to his fields in the evening "to say good-bye." As we stood next to the truck, the farmer turned and took in the view of dozens of towers.

"It's just ugly," he told me. "I long for the days when it was wide open country."

An hour later, all of it walking in hot wind and strong sun, I opened the front door of the Pattern Energy field office. The lone building had loomed against the bright sky for miles as I approached. There were several white pickups in the parking lot, but no people, so I had taken a moment to change my shirt, but I could not conceal my sweat. When I stepped into the office and the door swung shut, I felt a vacuum. Scott Creech, field manager for Pattern, strode up in jeans and a collared shirt and showed me the way to a conference room. Before we got to talking, I asked to use the bathroom.

A safety poster above the toilet showed a range of shades meant to represent urine. The darkest indicated potential dehydration. I had left the Bichsels with more than a gallon of water and had been drinking regularly, so I was in the clear. But the sign served as a reminder to mechanics who left the air-conditioned office to climb inside towers and tend to turbines that arid wind can wear a person down.

Back in the conference room, I sat in a cushioned seat, and Scott handed me a bottle of cold water. He explained how he had happened into the industry seven years earlier, when he was hired on from a natural-gas plant to work at a wind farm near Sweetwater, two hundred miles southeast. That area had been in the midst of a wind-energy bonanza, and Scott followed wind work to Nevada and then back to the Panhandle. Pattern, which operates more than a dozen industrial wind farms, from New Mexico to Missouri, Canada to Chile, had spent much of the past five years bringing Panhandle Wind I and Panhandle Wind II into Carson County.

Panhandle Wind I has 118 turbines on a total of fifty-two different parcels of land, with each turbine, built by General Electric, generating up to 1.85 megawatts of power. Panhandle Wind II has

79 turbines, built by Siemens, on forty parcels of land, generating 2.3 megawatts. Much of the Panhandle Wind I power has been sold in a thirteen-year deal to an energy hedge fund managed by an affiliate of Citibank. Much of the Panhandle Wind II power was sold to another fund run by a Morgan Stanley affiliate. The balance for each gets sold off at market prices into the ERCOT grid. All told, when the wind is blowing steady, the two farms can generate enough energy for more than fifty thousand homes, and the key for Scott and others is making sure the towers maximize the wind that blows their way.

Scott explained the clicking I'd heard atop a tower two mornings earlier, when I'd camped in Marvin Urbanczyk's cotton field. Fiber-optic cables running from the operation center to each tower help to deliver computerized commands that keep towers facing into the wind, as well as ensure turbines are not too hot or too cold. The goal is to have everything lined up when the wind is blowing twenty to twenty-five miles per hour 240 feet above the ground. Technical problems usually arise with big wind shifts or no wind at all. "As long as they're running," Scott said of the turbines, "they just keep running."

The Panhandle surge in wind came as a race against the clock, as the federal tax incentive for wind energy was winding down. The law, created in 1992, had spurred much early development in Texas and other key wind states. But it was being phased out between 2015 and 2019, and Pattern and other companies knew they had to fine-tune their systems to win investors for future projects. The continued momentum of industrial wind would soon be left to compete with cheaper fossil-fuel prices on the open market. "Everybody wants to feel good about producing clean energy," Scott told me. "But at the end of the day, you're dealing with bankers, lawyers, businessmen. Can you make more money doing this than investing somewhere else?"

As Scott spoke, he sat with his back to a large color map that covered the wall behind him. The map showed a satellite view of the northern portion of Carson County, with each tower and transmission line, multiple substations, and the Pattern field office. From the aerial view such elements merged into the terrain, as though ridges and rivers in the landscape. I was drawn to a small inset map, a rectangle about a foot or two wide and high, in the middle of the larger map. That inset showed a much larger geographic area: the entire Panhandle and neighboring states. I was struck by how much more land could be covered by more towers and transmission lines, essentially turning the center of the country—as well as windy spots around its edges—toward the harvest of the currents.

The farmer I'd met that morning was right that towers destroyed the lay of the land, in the sense of how people have long known the nature around them. But the greater destruction of climate change that wind farms help to avoid could far outweigh the loss of wide-open terrain. The turning of turbines as far as the eye can see could serve as a constant reminder of what humans have made of the world. And those towers would loom as a tangible sign of a more sustainable compromise to keep it all going. Had it come to this: Does preserving the natural order of the planet with renewable, carbon-free fuels require that we turn so much more of it over to fueling our existence? Are we destined, even with renewable energy, to claim more landscapes so the climate has a chance to keep its natural course?

I said good-bye to Scott and stepped into the gusty afternoon for two more miles of walking. I had a boost of energy from my break inside the building, but the temperature was still near ninety degrees, and the pack weighed heavy on my back. My steps fell with the steady repetition of a windmill turning, but there was labor in my legs, as the wind still hit me head on.

The road shot straight west one mile, then I turned south for another. I was aiming for a cluster of buildings and trees that I

suspected marked the family farm of Allene Stovall. The Bichsels
had called ahead to Allene, an old acquaintance. She lived in the
city of Panhandle, six miles beyond the farm, but she was planning to drive out and check on calves at the farm that evening,
and she told the Bichsels I was welcome to camp there.

Particularly difficult stretches of my walks, such as those last
two miles to Allene's, had become a favorite of mine. It is rare, in
this industrial world of constant communication and mechanized
mobility, to sense true solitude and the self-reliance of effort. As
an individual, moving step-by-step, I felt reconnected to place,
no longer separated from the source of things that fueled my life.
Moving on foot, I had overcome the numbing sense of dislocation
that had driven me out the door of my New Hampshire home,
three years before.

In such moments at the end of the day, I found psychological
strength. But I was still overwhelmed by the larger challenge of
responding to climate change. Even this deep into my Texas trek,
I was far from convinced that modern society—individual people
and the politicians they elect to make policy—would commit to
moving away from fossil fuels and embracing the power of water,
wind, and sun more completely. Yet I had a strong sense those
fuels could do the job. And I was eager to see more. A few months
after my Texas walk, I would make my last, across a desert valley
in California that is home to the largest solar power plant in the
country.

As I arrived at the driveway that led into Allene's farmyard
and the shade of trees, I was surprised to see the main house had
some windows broken and others boarded up. Shingles on a shed
roof had caved in. A small wooden tool building had the door broken off. All around equipment had been left to rest, no doubt for
years. An old crop sprayer, only twenty feet across, with slender
steel arms that reached out with hoses, had faded yellow paint.
Grass grew high around a John Deere Turbo 7720 tractor. A rusted

Chevy pickup sagged in the driveway. A dog beneath it roused, barking, and began walking my way. I slowed and talked to it quietly, and the dog slumped back into the shade.

It was only four o'clock, and I had hours to wait before Allene arrived to check on her calves in a neighboring pasture. With grass grown up against the buildings in the farmyard, I was on high alert for rattlesnakes waking from the long winter. I claimed a spot of shade on a small patch of mowed lawn near the tool shed, and I set my pack down. I leaned against it in a trance state, too tired to do anything, yet too alert to drift off to sleep. The only new thing in the farmyard was a shiny steel flagpole. Two crisp flags, one of the United States and another of Texas, hung at the top, fifteen feet off the ground. The flags snapped back and forth, their colors colliding, as the wind held steady.

Three hours passed, me leaning back in the ever-growing shade, the dog still sleeping in its oasis beneath the truck. The flags kept watch above us. Every half hour or so, a pickup sped past on the east-west road or a cargo train rumbled several miles south, the echo carrying across the soon-to-be-sown fields. Finally, a truck slowed and turned, rolling up the driveway. It was beaten up inside and out, but it slowed to a stop, and behind the wheel sat the bright smile of Allene Stovall, age eighty-three.

Allene had been born and raised on another farm, then spent years as a college sports pioneer, starting the women's basketball and volleyball programs at West Texas State. University administrators had warned her: if she wanted to keep teaching, she better stop promoting women's sports. She didn't stop, and decades later she was hailed as a leader of west Texas women's athletics. The women's basketball team at West Texas A&M, as the university is now called, would fly to Philadelphia to compete in the Elite Eight Tournament a few days after I met Allene.

She climbed out of the truck and showed me a garden hose connected to an old well, saying it would make good drinking water for that night. Then she cleared a space on the cluttered bench seat of the pickup truck and told me to climb in. As Allene drove slowly along the farm boundary, the dog, a stray she'd taken in a couple of years before, loped along behind.

Allene wanted first to check some irrigation lines in a field and soon spotted a drain cover that hadn't been properly sealed. She parked and took out a big wrench and gave the drain a few whacks. Tired, she handed the wrench to me, and together we finished the work. We climbed back in, and Allene drove toward the center of the property and the corrals holding the calves.

Allene talked about wind towers on her land that bring in $18,000 a year in royalties, which she appreciates. She said it took a while, but now the sight of hundreds of towers stretching across the horizon seemed as natural as newborn calves in spring. There had been one particularly hard moment, though. Several towers were built on the farm where Allene had grown up, several miles east of where we were. Power lines transporting the electricity hung too close to the abandoned farmhouse, and the company building the lines had recommended burning the house so that it wouldn't catch fire later. Allene had given permission and gone to watch the house burn. But when flames ignited her childhood home, Allene told me, she could not bear to look, and she drove back to town.

As we talked, Allene navigated her truck along the perimeter of hundreds of acres of pasture, then followed a road to the center of it. She stopped alongside a sturdy barn. She positioned the pickup so we had a view through a fence and down a slope into a corral straddling both sides of a stream. A dozen or so cows were in the corral, some with calves nearby, and three men—two on ATVs, one on foot—were trying to get ahold of one cow that had begun to give birth. It wasn't clear why the men needed to interfere, but the cow seemed bothered by their company, and she jogged one way and another, as the men circled in.

Just upstream, a low dam, built by Allene's father decades earlier, held back a small pond, and songbirds flitted through the pasture. The day's heat had broken in the twilight, and the men moved fast. One of the cowboys, named Melvin, worked a rope. The pregnant cow, more nimble on its feet than I'd expected, continued to bolt, even though the calf's head and front hooves had emerged. Another of the men lunged and grabbed the calf's front legs, to help it along. The cow bucked, running forward, pulling the man onto the ground, his arms now up in the air, still holding the calf, which popped all the way out and landed on the ground too. Allene and I watched the action, and as the calf caught its first breaths, she said, "Calves are so durable. Oh my land, so durable. That one was born going fifteen miles per hour."

The cow had already jumped across the creek and was standing still, deep breaths heaving her sides up and down. The men wanted to make sure she nursed the newborn rather than abandon it. They tried to maneuver things so nature could take its course. One ATV sped up the hill above the cow and angled back, encouraging her to jump the creek toward her calf. Another moved in from the other side. The cow did not budge. Melvin nudged the calf with his boot, urging it upward, hoping it would stagger in the cow's direction. The cow seemed disinterested in the calf for a full five minutes. But then, overwhelmed by instinct or relenting to the designs of the men around her, she strolled over to the teetering newborn. She licked its slick coat for several minutes, then stood stock still as the calf found her udder and drank deeply.

Allene and I remained more or less silent. The sun had started to sink on the horizon, and its glow washed the scene against time. Hearty gusts grew stronger as the calf settled on the ground for its first night in an uncertain world. Allene, without prologue or grand proclamation, shared a lesson from eight decades of experiencing change on the wind-whipped terrain. "It takes a lot of patience," she said.

# Chapter 6
# Whether to Burn

During those final minutes descending toward the desert, passengers sat in an air-conditioned daze, idly turning silenced iPhones in their hands in anticipation of the moment they would rejoin the terrestrial flow. Beyond the airplane windows, the land was crag and dust, ridge and canyon, gullies washed by water through millennia but not on that day. All in the desert basin was dry earth, red and yellow and brown, but mostly brown, charred, like the inside of a brick oven. The plane passed another jagged ridge, and it looked as though no person could ever survive such a place. I would soon begin my final walk: through the rugged landscape and the country's largest solar power plant. In that intense isolation, might some kind of solution wait?

I had first to navigate the industrial maze of Las Vegas, and as the plane passed above another ridge of red rock, tracts of houses came into view, quick and complete: whole city blocks of homes clustered tight. Stout, short driveways led to tiled roofs framed by patches of bright green lawn and the glistening blue of backyard pools. This was the east side of the city, and the plane passed over McMansions not ten years old, then blocks of more modest bungalows from an earlier era. After only another minute, the plane banked south, and out the window to the west spiked towers of gold and silver and black. On that June day in 2017, three months after I'd walked across the windy plains of Texas, the hotels and casinos of the Las Vegas Strip shielded and shone, as though insulated extensions of a subterranean world reaching up into the heat.

Las Vegas's unprecedented bloom in the desert had mirrored the expansion of the fossil-fueled industrial age through the twentieth century and into this one, with the city's manufactured landscape of endless entertainment rising from an outpost of a few dozen people in 1900 to a metropolis of roughly two million residents by 2010. Yet I, having slept next to oil wells, a natural-gas compressor, and train tracks delivering ever more coal, could see in the shimmering city outside the airplane window only a symbol of the consumption that made the places I had walked such urgent terrain.

Despite decades of data showing that too much carbon dioxide was entering the atmosphere, more than 80 percent of all energy consumed was still being fueled by oil, gas, and coal, burning through deep time at expense to the civilization that craved it. The political reversal just months earlier that brought Donald J. Trump into the White House ensured that the United States, at least, would not change course anytime soon. Trump encouraged more fracking for oil and gas and rolled back a ban to extend the harvest of Powder River Basin coal, touting the black rock as a fuel for the future. The U.S. Energy Information Administration was already predicting that the nation's energy consumption could grow more than 11 percent by 2040, and in that year, more than two decades away, fossil fuels would still provide more than 75 percent of all energy. Renewable fuels would grow faster, the EIA stated, than any other source of energy. But that burst was coming from such a small base—solar had contributed less than 1 percent of all electricity in 2016, for example—that water, wind, and sun power combined would likely account for just 10 percent of all energy in 2040.

The transition to a renewable, carbon-free future, in other words, was not happening fast enough. And the danger was only accelerating: the amount of carbon in the atmosphere—at 405 parts per million by 2016—had jumped more in that and the previous year than in any other two-year period since measurements began in 1956 at the National Oceanic and Atmospheric Administration's Mauna Loa Observatory. The year 2016 marked the fifth consecutive year that carbon dioxide in the atmosphere had risen by more than two parts per million, another record. Pieter Tans, a scientist with NOAA's Global Greenhouse Gas Reference Network, stated in an announcement of the measurements two months before I arrived in California, "The rate of $CO_2$ growth over the last decade is 100 to 200 times faster than what the Earth experienced during the transition from the last Ice Age. This is a real shock to the atmosphere."

As I looked out at the silver city beneath me, I saw in the skyscrapers not feats of engineering that made the desert habitable, but fragile towers tempting the collapse of so much that had been built. Still, my last walk would offer promise, exploring an alternative closest to the source: the rays of the sun. My route would traverse the Ivanpah Solar Electric Generating System, where hundreds of thousands of mirrors were redirecting sunlight toward a tower that converts its energy into electricity. This journey on foot would be my shortest, just ten miles up the steep slope of a valley fifty miles southwest of Las Vegas. But it would also be the most physically daunting, with hundred-degree heat much of the day.

The plane set down on the concrete runway of Las Vegas's Mc-Carran International Airport, rubber tires skipping and settling as the big steel tube bumped along. It was two in the afternoon, and the temperature outside measured 105 degrees. A voice came over the intercom, instructing passengers to turn on overhead air vents, turn off reading lights, and pull down window shades, lest the plane overheat as it waited at the gate to take off again.

I shuffled down a corridor and into a wide hall, where first my backpack and then my walking stick emerged on a conveyor belt. The stick, so far removed from the North Dakota prairie where I'd found it on the first day of my first walk, three years before, had been damaged on the cross-country flight. A split in the wood measured ten inches down one side of the stick. I ran my thumb along the crevasse, then set the end of the stick on the slick floor of the baggage claim area. I leaned my weight against it. The stick was still sturdy enough to support me.

I climbed into a rental car and spun the steering wheel toward the concrete canyons of the city, navigating between high walls built to dampen noise for those living on the other side. I hovered at a stoplight alongside other people in other air-conditioned cars. My day had begun before sunrise at home in New Hampshire, and as night fell I stopped at an REI for maps, fuel for my camp stove, and a compass. I picked up fresh tacos and cold beer, then drove south among the sprawl of Henderson, a second city blending seamlessly into the edge of Las Vegas. There I checked into a Hampton Inn and a room chilled to sixty-eight degrees.

I sat on the edge of the bed, gulping *carnitas* and turning channels. A weatherman stood before a map colored orange and red and warned locals the temperature would hit triple digits all week, the first extended run of such heat that spring. "Be careful out there," he said.

I hoped that by moving on foot through the Mojave Desert in the punishing heat of June I would find confidence in a renewable fuel that could help power the energy system and in a personal experience that could buoy me going forward, no matter the trends of today or the temptations of tomorrow to keep consuming fossil fuels. Each of my earlier walks had brought me closer to nature, and there I had found strength. But that evening I was scared. I had been having dead-of-night dreams for weeks back home in New Hampshire, worrying about the heat of the desert and its Mojave Green rattlesnakes, the most venomous in North America and native to my route. Did I really need to expose myself to nature this last time, so far from the predictable safety I shared with my family?

I wondered in that moment, as I often had traveling toward earlier walks, of a bigger problem: was it too late to find another way forward? I was such a product of, and participant in, the industrial world. Just as I felt nervous about leaving the shelter of the hotel, was I, and everyone else, destined to be overinsulated from change by an energy system too big to control or correct?

Despite raw nerves, I slept soundly in the cool, cotton cocoon of the hotel's bed, then rose early to eat oatmeal in the lobby lounge. The temperature outside was already near ninety degrees, and I wandered out the front door with a cup of coffee. The sun beat in a blue sky, and the wind was so dry it seemed to scratch my bare arms. I made it only a few hundred yards before I angled off the asphalt parking lot into the shade of a palm tree. It was a quarter after eight.

※

The sun is one of trillions of stars in the universe, but closer by far than others to earth, and its intense gravity keeps the planet in its orbit. That gravity also forces the massive ball of gas in on itself, with hydrogen at the sun's core converting to helium, a nuclear reaction that sends energy outward. For 4.5 billion years solar radiation has escaped into space, sending light and heat toward earth. That energy has created life on the planet, fueling the growth of plants, the basis of the food chain, and creating conditions for the atmosphere to exist.

The nuclear reactions at the sun's core are perpetual, and the star is thought to have enough gas to last another 6 billion years. In the meantime, the orbit of the earth ensures the sun's rays arrive everywhere on the planet, every day. The intensity and length of those rays change from one place to another, depending on many variables, from the angle of the earth's rotation to the cover of clouds. But to some degree the energy of the bright star arrives, and as early as two thousand years ago humans used magnifying glasses to start fires. Early cultures around the world faced dwellings toward the sun to capture daytime heat. But it wasn't until roughly two hundred years ago that scientists discovered the sun's rays could be converted to electricity. That potential was little more than a curiosity for much of a century, as oil, gas, and coal burned ever brighter. But government and commercial efforts to harness the sun accelerated in the 1970s, with the global oil crisis, and have continued with the challenge of climate change.

The sun's energy is measured in kilowatt hours per meter squared per day. In the United States no place receives such intense insolation, as it is called, as the desert Southwest. A swath of earth that traces the Mojave Desert through southern California and into Mexico receives the most sun of all. The Ivanpah Valley is one small piece of that sun-scorched terrain, its western slope in the path of particularly strong arriving rays. The heat that comes with them has defined the landscape in the valley and surrounding desert for millions of years, emptying lakes and baking rock, with plants and animals adapting to exist in a place that is most often arid, strikingly hot in summer, bitingly cold in winter.

From Las Vegas I drove toward the California state line on the wide lanes of Interstate 15 and pulled off in Primm. The strip of casinos, hotels, and restaurants sits on the Nevada side of the border, and, just beyond, the long, shallow arc of Ivanpah Valley runs southwest into California. The roadside oasis offers some kind of comfort to people making the 270-mile crossing between the settled structures of Los Angeles and Las Vegas. I parked near dozens of gas pumps and followed other travelers into a convenience store. We staggered up and down aisles, collecting big cups of soda and sandwiches packed in plastic. We paid with credit cards, saying little to one another, temporarily suspended in a world that did not suggest we stay.

I drove to the edge of Primm, on higher ground, and I had my first distant view of the Ivanpah solar facility: more than three hundred thousand mirrors across six miles of earth. The mirrors formed a blue-gray sea of light as they reflected the sun's rays toward each of three high towers, which appeared, despite the extraterrestrial dimensions of their design, to nurture.

I merged back onto Interstate 15, crossing into California and the Ivanpah Valley, framed on both sides by sculpted brown peaks. Along the shallow floor stretched Ivanpah Dry Lake, a cracked, dusty bed two miles wide and twenty miles long. Archaeological remains indicate humans moved through the valley as early as twelve thousand years ago, when the lake still held water. Speeding alongside it at eighty miles per hour, I had the sensation of being in the trough of a vast wave, the mountains cresting on each side of the valley.

The dry lake and all around it had been shaped during millennia by floods and eruptions, erosion and evolution, and I would learn that even the distant dark slopes were home to colorful cacti and grass, jackrabbits and tortoises, each finding sustenance to survive. Interstate 15 began a long climb up the west slope of the valley toward the peak of Clark Mountain, and I exited the highway and turned toward the valley's eastern edge. I could see a shadowed oasis eight miles in the dusty distance, and as I approached, its edges took shape as trees and trailers and low buildings. I crossed two sets of train tracks, first built in the early 1900s and still run by Union Pacific, and pulled into a mostly empty lot. An adobe building had letters on its facade that announced "Nipton Hotel."

The hotel had five bedrooms and a common room with couches. I took room number 3, which had been a favorite of Clara Bow, the silent-movie star, who had enjoyed escaping Hollywood to the desert outpost eighty years earlier. The afternoon I arrived, a half-dozen organizers of the Mojave Death Race were shuffling around the common room, packing up boxes of hydration mix and empty coolers. The race, a 250-mile relay on foot and bicycle across the desert, had finished that morning. The organizers were whipped and weary as they hurried back to busy lives around Los Angeles, and soon I sat alone on the hotel porch in soft evening light, looking across the valley at the western slope, where I would walk.

I would need to worry about water as I hiked to the innovative solar plant and about isolation as I camped among the cactus canyons beneath Clark Mountain, three thousand feet of elevation above the dry lake. Even with all my experience on foot, from the North Dakota plains to Wyoming grasslands, from the coast of Maine to the Texas prairie, this last walk overwhelmed. Alone and exposed, I would take one last measure of my relationship with nature.

During the years of my walks across landscapes of fuel, the industrial system of which I am a part appeared only to be accelerating its overheating of the earth. That March, as I had been walking through the hopeful winds of west Texas, scientists were proofing reports that documented global warming is occurring more quickly than previously thought. The most dire climate predictions contend that sea levels may rise six or seven feet by the year 2100: roughly eighty years away. The same amount of time elapsed between my grandmother's childhood and my adulthood. In another eighty years my children's future children would be adults. Will humans find answers to the climate challenge in time?

Sitting on the hotel porch that first evening, an ice-cold Estrella beer in my hand, I watched the sun drop behind the ridge of the Clark Mountain Range, and the shifting light dramatically changed the scale of the valley. The sandy brown bed of the dry lake, six miles away, seemed to be floating closer, and the mirrors of the Ivanpah solar facility on the slope beyond settled in a soft gray, as though the lake had moved higher up. The sinking sun would soon be gone for ten hours, until it would rise again over the New York Mountains east. The towers at the center of the field of mirrors began to cool, the bright white blocks at the top shining silver.

Such a vast operation had not come easily into Ivanpah Valley. When a company called BrightSource announced the idea, environmentalists opposed putting an industrial plant adjacent to the Mojave National Preserve, protected desert terrain stretching southward from the valley for more than fifty miles. The six miles of slope where the solar facility was sited is home to desert tortoises, an animal threatened with extinction. There also was concern because the towers operate at such high heat that birds passing close by explode in flight. As with wind turbines, was the solar plant and its carbon-free fuel a compromise worth making?

The benefits from attempting such an unprecedented system won the day: enough energy to power seething cities and lessons learned in construction and engineering to help build better solar plants elsewhere. During the Obama administration, a federal program guaranteed $1.6 billion in loans for the construction of the system, and the Ivanpah Valley, with Primm at one end and Interstate 15 splitting it down the center, was deemed to be a place already so altered that a solar facility would not make too big a difference.

Since 2013, then, the mirrors have tilted to follow the sun across the sky and send its rays to the top of the towers, which become orbs of energy. Some travelers on Interstate 15 pull off the exit at Yates Well Road and stand, phones in hand, taking pictures. Pilots passing one hundred miles away can see the mirrors glowing on the desert floor.

From my perch on the hotel porch, the solar plant all but disappeared in the dark of night. Gusts of wind still swept across the dry lake, rocking cottonwoods that surrounded the adobe hotel. Along the distant ribbon of Interstate 15, lights of cars and trucks, ever moving, continued across the valley toward the pass near Clark Mountain. I thought of so many rhythms at once: a driver peering through the windshield of a hurtling car, a snake slithering out of a hole in the ground, the branches of a creosote bush bending in the breeze, the sun on the far side of the spinning earth. After that rotation ran its course, the sun would rise high and hot across the valley again, amplifying every detail in bright light and stretching again the range of sight.

By eleven o'clock the next day, I was on foot in the center of the valley, walking up the western slope from the dry lake and Interstate 15 toward the Ivanpah solar facility. I carried my pack on my back and stick in hand. My mouth was dry from only a short effort. The terrain, at more than three thousand feet above sea level, was already climbing steadily. I wore protective gaiters on my legs in case I came across a snake, and I carried more than a gallon of water in my pack. The temperature measured 103 degrees, and I tried to keep long strides.

More than a mile from the main gate, mirrors already surrounded me in orderly rows. The mirrors were mounted atop steel poles in sets of two, each set six or seven feet above ground. A white pickup truck sped past toward Interstate 15. I kept walking alongside the asphalt road. More chain-link fence on each side guarded more rows of mirrors. The road made a sharp left turn, and I had the sensation of weaving into a force field, stealthily approaching the center of an unknown organism. After twenty minutes I arrived at a cluster of low buildings, where I stopped at a security gate meant for vehicles. Once cleared, I would be deeper inside the machine that was capturing the sun's power to fuel distant lives. The wind whipped from the south, adding urgency. I pushed a button and waited.

⁂

Near the center of the third field of mirrors, a heliostat began to pivot with the shifting sun. The heliostat, which takes its name from Greek words meaning, essentially, stationary sun, was sleek and sturdy, its anchored pole holding a frame with two mirrors on top. Wires delivered data from software that tracks the location of the sun, and on that June afternoon, in that moment, a small motor on the frame of the heliostat reacted to the data with soft clicks and a steady whir, and the frame turned only an inch, if that, the flat-panel mirrors turning, too. All this happened so subtly that it was almost imperceptible from where I stood a few feet below the mirrors, their shade giving me cover.

The rays hitting the mirrors overhead followed the new trajectory toward the tower, which stood 450-feet high at the center of the field. The heliostat was one of more than fifty thousand heliostats arranged in orderly arcs, together pointing more than a hundred thousand mirrors at the lone tower. This concentration of the midday sunrays hit panels near the top of the tower, and those glowed white hot, with an intensity that indicated a great depth. From my position on the ground below, the panels seemed to pulse, but it hurt my eyes to stare for longer than a few seconds. Still, I could not look away; the power of the panels absorbed me, as though they may lift me into another dimension.

Behind the panels, stout pipes delivered water to the top of the tower, where it was heated by the concentrated sunlight. That

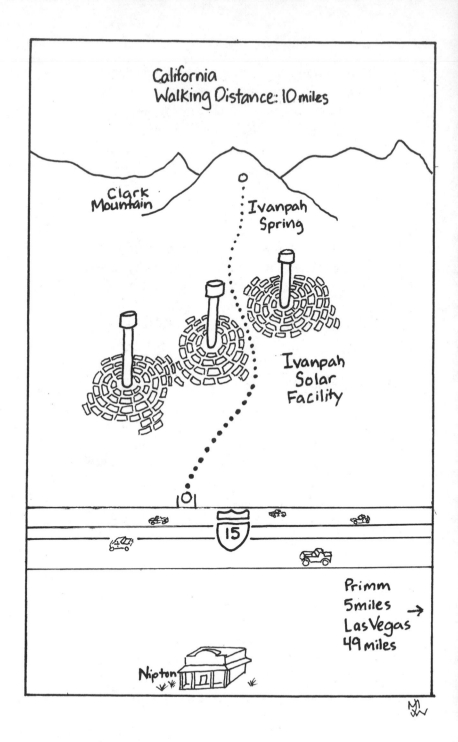

water descended toward a generator, then rose a second time through more pipes to be superheated above a thousand degrees, before descending again to create steam to drive a turbine. The turbine was housed in a long building at the base of the tower, and as it turned it generated electricity, which passed into wires, which connected to others.

To the south of where I stood among the heliostats, a second field of more than fifty thousand heliostats, each also holding two mirrors, heated a second tower. And beyond that, a third field heated a third tower. Despite the immensity of the operation— more than three times the steel used in the Eiffel Tower to build the heliostats and towers, enough glass in the mirrors to replace all the windows in the Empire State Building fifty-four times— its work was deceptively simple: generating enough heat to create steam to turn a turbine. That had long been done in power plants fired by gas and coal and nuclear reactors. Skeptics have dismissed the solar field as much work for little result. "We've just technologically advanced ways to heat water to make steam," one man told me before I set out walking.

But as I stood among the mirrors, I could see that the machine all around me held an essential promise for the future: one part of an attempt to reengineer the industrial system so that it does not change the climate so much and so severely threaten life as humans have come to know it. The solar plant's elemental simplicity was its strength. And in the Ivanpah Valley, at the early edge of the twenty-first century, those thousands of mirrors were capturing many times more of the sun's power than had ever been done before by a concentrated solar plant. The energy created traveled in wires up the rugged slope of the Ivanpah Valley, over the sharp ridges of the Clark Mountain Range, and hundreds of miles across the Mojave Desert toward the cities of southern California. Two towers fed customers served by Pacific Gas and Electric. The third tower fed customers of Southern California Edison. All told, the mirrors redirected enough sunlight to heat water to create steam to send electricity to more than 140,000 households.

In the low buildings near the main entrance, while the helio-stats clicked and whirred to keep pace with the sun, a half-dozen people in an air-conditioned, windowless room monitored the machine. They sat behind desks arranged in arcs and faced a wall with dozens of flat-panel screens showing the status of the pipe through which water traveled and of the wire that carried the energy created from the steam. The maneuvering of all of it was controlled by elaborate software—called SFINCS, or Solar Field Integrated Control System—that determined the angles of the heliostats, the temperature of the boilers, and the spin of the tur-bines. Maintenance crews shuttled to specific points among the mirrors and towers when a problem occurred, and a big red but-ton near the center of the room offered help in the event of an ex-treme emergency: with a push of that button, all of the more than three hundred thousand mirrors would be redirected to take the sun's rays off the towers within twenty seconds.

I had negotiated access into the Ivanpah solar facility on the condition that I not name or quote anyone with whom I spoke about how the system worked. That was fine with me, as this far into my journey I was not interested in particular people so much as what their collective effort represented: power that offered a new way forward.

Such massive movement of mirrors, and the hot-water heating system it fueled, is just one form of solar power and not nearly as common as the photovoltaic panels that have been increasingly installed on residential rooftops and in vast fields for industrial-scale generation. Silicon cells within those photovoltaic panels convert sunlight directly into energy. One industrial field, more than a mile square, sprawled just south of Primm, as large as the city itself. Another sat just east of the Ivanpah concentrated solar system.

But for areas of particularly strong insolation, such as the desert Southwest, the Ivanpah solar facility and its concentrated solar system offered potential to generate even more heat. As I had been plotting my last walk that spring, I'd been drawn to its tangible design of mirrors and towers and turbines, in which the sun's rays are redirected to create heat, a process familiar to any child who has turned a mirror at a crisp leaf and waited.

The Ivanpah solar facility came with one key compromise. Natural-gas boilers, which emitted carbon into the atmosphere as they worked, began heating the water in the pipes for an hour before the sun rose and for an hour after it set, helping to keep the temperature high enough to make the most steam from the sunlight as it arrived. Engineers hoped to make the system so efficient that it wouldn't need the natural-gas boost. But even using some natural gas, the operation did not have the feel of chaos that came from the opened earth that I'd seen in North Dakota, Pennsylvania, and Wyoming.

I walked from the heliostats to a point directly beneath the base of the tower, where the descending pipes ran several stories above the ground. After the water passed through the generator to make the steam, it traveled through an air-cooled condenser next to the tower. That was also raised off the ground, and I stood in the shade beneath fans twenty-five-feet across.

It was an odd sensation, idling beneath all those spinning blades, in shade from the steel structure above me, the industrial machine anchored atop the natural world. Wind and hot air rushed all about, but no odors or explosions, no underground leaks and so much less carbon burning as the fans turned overhead. For a full ten minutes I had the place to myself, and I looked up at the hot-water pipes, wondering. Could this design help save the industrial system: electricity generated from existing currents rather than dug up from deep time?

Engineers at the Ivanpah solar facility had learned in the first three years that high cloud cover intercepts more of the sun's strength than they had anticipated before it was built. They focused efforts on maximizing the use of the sun as it arrived during the day, with up to fourteen hours of operation during summer but only eight in winter.

And they chafed against the biggest challenge of renewable fuels: storing excess energy at peak moments of production so that it can be used when people demand it. Improvements in industrial-scale batteries will help photovoltaic solar plants, as they will with wind power. But concentrated solar has another hopeful option. The Crescent Dunes Solar Energy Project, another concentrated solar field in central Nevada, was using a system of salt-water storage, in which the sun's energy heats not water but molten salt to more than a thousand degrees. That block of salt can keep the heat for up to ten hours, an available source to create steam and convert to electricity as the energy grid requires it. The system, launched in 2016, is one piece of the race to improve technology and reduce costs. The U.S. Department of Energy predicts one ambitious target may be hit: concentrated solar plants could produce electricity by 2020 for six cents per kilowatt hour, a price competitive with other fuels.

As I stood beneath the tower on that afternoon in 2017, the Ivanpah solar facility was already showing much of the potential for power that can come from wanting to engineer systems to harness the earth's existing energy. As water in the pipes ran up and down the tower, then passed through the condenser above me, I could feel the living energy: a transfer of power from sun to steam to electrons far more sustainable than unlocking long-buried carbon and sending its emissions into the atmosphere. For parts of the world with intense insolation, the potential for capturing massive quantities of power still may come most from concentrated solar systems, and the Ivanpah solar facility was both reality for today and experimentation for tomorrow.

By three o'clock I had walked from beneath the tower back among the heliostats. Again the motors whirred and clicked, with the mirrors shifting slightly to match the sun's movement across the sky. The still-strong rays angled toward the tower, heating the water to more than a thousand degrees. The power lines running from the solar plant crackled as the voltage began to travel up and over the mountain ridge.

※

A half mile west of the solar plant, back in the open desert, the slope of the valley rose even more steeply toward the Clark Mountain Range, and the road surface switched from asphalt to dirt. The brim of a white hat hung low over my eyes. It was four o'clock, and the sun would soon begin to relent, but not yet, and dry air whipped up from the south, the temperature still above a hundred degrees.

I stopped after a few more minutes of walking and stared at the broad spike of Clark Mountain, at 7,933 feet the highest point in the Mojave Desert. I planned that afternoon to follow the dirt track toward it, climbing 1,300 feet of elevation over four particularly intense miles. The ground beneath my feet held many layers, with gneiss, migmatite, and other rock 1.7 billion years old. Magma flows had given way to rising mountains, which later sunk beneath an inland sea. The ocean withdrew and more volcanoes came, with tectonic shifting before and after pushing up ridges.

Along my route fossils had been found of multicellular organisms, such as mollusks, but also single-cell creatures that had lived long before. In the thousands of years since the Ivanpah Dry Lake drained, native peoples traversed the desert toward the coast and back or climbed high into the mountains, along more or less the same path as mine, in the white heat of summer. In recent centuries prospectors had scoured the slopes of the Clark Mountain Range for unusually rich deposits of silver, copper, and gold, but only traces of that life remained by the day I was walking up toward those heights.

After thirty minutes more I stopped and leaned with the weight of my pack on my stick. I was already several hundred feet of elevation above the solar field. Behind me the rocky road ran back toward the third tower, beneath which I had stood just an hour before. Plants two, three, and four feet high had claimed perches in red and brown earth around me. Flat paddles of prickly pear, dusty green with rusty edges, huddled above packed dirt. Spindly spikes of a well-branched cactus rose up from between a cluster of rocks, arcing above a bright green bush with white blossoms. The green of another cactus was overwhelmed by the white of longer spikes, and the sharp fronds of an agave fanned above fat trunks.

I had relished such moments during earlier walks, as more than encountering any particular source of energy, my odyssey on foot had become about embracing exposure: living in nature, unsure of what would come next. Traversing the North Dakota oil field and weaving among open-pit coal mines, particularly, but even striding between towns in the wind-struck Panhandle of Texas, my steps had often brought unexpected encounters with people. On this final walk in the desert, I knew that I would face solitude. Off-road adventurers sometimes drove all-terrain vehicles along the dirt road I walked, exploring the remains of old mines—but not in June, when temperatures rose to their peaks, and the sun dominated long days. No one else would climb the heights of the remote reach of desert on this day.

A jackrabbit bolted among cacti and creosote. Snakes surely lurked, but in that high heat of afternoon they would be deep in cover. The Mojave Green, a plump, short rattlesnake known for its aggressiveness, delivers two toxins when it bites. One toxin attacks the nervous system, another the circulatory. I set down my pack. The odds of encountering a snake were low. But somewhere around me a snake surely was coiled in the shade of creosote and mesquite bushes. A snake's scales blend into the palate of browns at ground level. It sits in wait, tongue emerging to gather scents to deliver to glands in its mouth. Pits next to its eyes measure the heat of nearby objects, in search of passing prey. A snake would not hunt me, of course, but an unintentional crossing could prove fatal, especially given my isolation. I smelled the wind. I bent and rolled rough pebbles between my fingers. Moving fast would increase the chance of an accident.

I carried in my pack a gallon of water, having filled up my bottles in the offices of the solar plant. I had only my stick by my side, but it had helped me keep balance since that first day walking on the North Dakota prairie. As I stood on the slope, eyes scanning the base of bushes for anything waiting, I found courage, moving from fear to a feeling of peace and belonging. In the bright light of the desert day, I was less afraid than I had been in the Hampton Inn a few nights before.

The surface of the earth was brittle, sedimentary layers of sand, silt, and limestone that shifted easily, redefined by my steps. I turned and surveyed again the valley sweeping beneath me. The sun had settled above the ridge to the west, and the photovoltaic panels near Primm, at the north end of the valley, darkened toward black. The six-mile field of mirrors at the Ivanpah solar facility had shifted color, from silver to bright blue, and the heating panels atop the towers also, from white to silver. Those towers, which had loomed so large when I stood beneath them, now seemed proportional.

During the days of my walk, researchers were preparing to publish a report in the journal *Nature Climate Change* that would underscore the immediate need for renewable-energy innovations. The team of authors included specialists in statistics, sociology, economics, and atmospheric science. Their research calculated that, given recent trends in fossil-fuel consumption and other social factors, the earth was 95 percent likely to warm more than

5.5 degrees Fahrenheit by 2100, well above the target threshold set by the Intergovernmental Panel on Climate Change.

I walked farther toward Clark Mountain, stopping every half mile or so, taking care to drink water and keep my heart rate steady. The contours of the valley softened in the angled light, changing again the scale of the place. A train ran past the Nipton Hotel, fifteen miles east, but I could not tell whether its cars held cargo or coal. Interstate 15 remained a ribbon of movement. I could hear only wind, an audible expression of heat, and it calmed the world, ever moving. The valley, with its ceaseless consumption and sources of solar power, seemed in that moment an example of the answer needed: humans and nature forming a new order.

A wide gully sculpted a slope toward higher ground, and the road I followed traversed a shallow wash. With the gentler rays of the diminishing day, I hung my hat on my shoulders. I soon would encounter more layers ahead, as a few hundred yards west lay ruins of a mining settlement. In the 1870s the town of Ivanpah had been home to hundreds of miners, with hotels and saloons, stores and a blacksmith shop, a post office, and more. Prospectors had arrived from over the mountains to the west, shipping back silver on a road cut through the desert to San Bernardino. Fortunes came quick and hundreds of miners harvested while the getting was good. But twenty years after the rush had begun, only eleven people lingered, as the price of silver was dropping, and the richest ore had already been taken.

I followed a narrower track toward an old water tank. Bushes stood higher than my shoulders. I set down my pack and continued uphill, angling toward a lush, green tree standing above a spring and the ruins of a mill site. With only my stick, I did begin to wonder what else was around me. Burros and mountain goats roamed the range, but also coyotes and cougars. Soon I stood facing timbers wedged in the side of a hill, the only visible remains of the mill. I could not see Ivanpah Valley, surrounded as I was by the higher vegetation, but I felt its pulse. I idled in front of the timbers, connected at once to past and future. The sun would drop in an hour behind Clark Mountain, and I would sleep in the present of a desert night. I held my walking stick in my hand and spun, vulnerable and alive.

I had known that sensation in the alfalfa field in North Dakota, as I lay in my tent and watched the leap of the flaming flare and listened to the rumble of trucks collecting oil from wells. And in Dimock, Pennsylvania, I felt it while perched on the hill on the Teels' farm, waking to the morning fog and the roar of combustion as gas compressed into underground pipelines. And again it was there, alongside train tracks in the Powder River Basin of Wyoming, as coal traveled toward distant markets. But my time beneath the sun at Ivanpah Spring also came with a sense of confidence, like that I had found on the surging coast of Maine and on the windy plains of Texas.

I walked back to the base of the gully and the open view of the valley, with its sprawling solar plant and congested highway, to make camp for one last night. I had set out from my basement three years earlier to overcome an intense dislocation from the source and find understanding there. My walks through terrains of fossil fuels and renewable, carbon-free energy delivered me to this simple demand: care enough to see ourselves as part of the planet. Such humility is essential for our urgent attempt to realign the industrial world on natural currents.

Would the insulation of my daily life at home overwhelm my awareness of that? The next afternoon I would begin heading back to New Hampshire, cars and airplanes hurrying me into that industrial world. But first, I would rise before dawn at Ivanpah Spring and hike a steep canyon four miles higher. The trail would clear a ridge, and I would lower my pack at the edge of a meadow, the peak of Clark Mountain looming. A hawk would soar on warming air, as the meadow, buffered by the ridge from Ivanpah Valley, rose to its own rhythms. I would watch wild burros graze in a steep draw and jackrabbits dart among their hidden homes. I would stand still, stick in hand, and take strength from the heat of the sun on my skin.

# Acknowledgments

Peg Hellendsaas, the owner of the campground at Tobacco Garden, said to me the evening before I took my first steps into the oil field of North Dakota, "I admire what you're doing." Such confidence in my uncharted journey carried me forward, and I am deeply grateful to the hundreds of people who guided me over the next three years, offering information or encouragement, cold water or a safe place to sleep. As I encountered new terrain, so many strangers quickly became friends.

At the University of New Hampshire, many colleagues gave crucial support, including, Michele Dillon, Burt Feintuch, Sue Hertz, Lisa Miller, Jaed Coffin, Andy Merton, Meg Heckman, Rachel Trubowitz, Dave Howland, and Scott Weintraub. UNH's Center for the Humanities and Carsey School of Public Policy provided time and money to head out the door.

Agent Paul Bresnick believed in my story from the beginning and took care of it until the very end. Joe Yonan and Neil Swidey helped me get started. Terry McDermott sharpened the focus early on. Alex Tizon, who had taught me two decades earlier to explore "multiple layers of reality" in every story, kept me close to the land on this one. Marla Williams dissected drafts with care and conviction, dedicated to helping me say what I hoped to. Essdras Suarez lay in desert dirt with a five-hundred-millimeter lens to capture the cover image. Matt Wimett created maps that chart the spirit of each place through which I walked. At ForeEdge/UPNE Stephen Hull, Susan Silver, Amanda Dupuis, and the production team expertly brought it all to bound life.

My parents have always nurtured curiosity. Julie inspires my search for stories. Luca makes sure I keep moving quickly. Colette shows me how to find my voice when I stop. I am at home with them.